SpringerBriefs in Computer Science

For further volumes:
http://www.springer.com/series/10028

Xiaohui Liang • Rongxing Lu • Xiaodong Lin
Xuemin (Sherman) Shen

Security and Privacy in Mobile Social Networks

 Springer

Xiaohui Liang
Department of Electrical and Computer
    Engineering
University of Waterloo
Waterloo, ON, Canada

Rongxing Lu
School of Electrical
    and Electronics Engineering
Nanyang Technological University
Singapore

Xiaodong Lin
Faculty of Business and Information
    Technology
University of Ontario Institute
    of Technology
Oshawa, ON, Canada

Xuemin (Sherman) Shen
Department of Electrical and Computer
    Engineering
University of Waterloo
Waterloo, ON, Canada

ISSN 2191-5768   ISSN 2191-5776 (electronic)
ISBN 978-1-4614-8856-9   ISBN 978-1-4614-8857-6 (eBook)
DOI 10.1007/978-1-4614-8857-6
Springer New York Heidelberg Dordrecht London

Library of Congress Control Number: 2013947689

Springer is part of Springer Science+Business Media (www.springer.com)

Preface

Social networking makes wireless communication technologies sharpening tools for extending the social circle of people. It has already become an integral part of our daily lives, enabling us to contact our friends and families without geographic barriers. Recently, social networking can be further reached in a mobile environment due to the pervasive use of smartphones. Smartphones greatly increase our communication, computation, and information storage capabilities and help us stay connected with our social friends anywhere anytime. The convergence of smartphones and social networking tools give us a pervasive communication platform, named mobile social network (MSN), over which we are able to use long-ranged wireless communication techniques such as 3G and LTE to reach our online social friends, or short-ranged techniques such as Bluetooth or WiFi to explore the physically close neighbors. In MSNs, multiple communication domains coexist, and within each domain promising applications are fostered. For example, nearby friend search applications help users to find other physically close peers who have similar interests and preferences; local vendors disseminate attractive service information to nearby users and users leave service reviews to vendors. Before the practical implementation of these applications, there are still many challenging research issues, among which security and privacy are the most important ones as users are vulnerable to security attacks, easily ignore their privacy protection, and hardly have trust toward others in the MSN. In this book, we focus on three emerging research topics in MSNs, namely privacy-preserving profile matching (PPM) protocols, privacy-preserving cooperative data forwarding (PDF) protocols, and trustworthy service evaluation (TSE) systems. The PPM helps two users compare their personal profiles without disclosing the profiles. The PDF helps users forward data to their friends via multiple cooperative relay peers while preserving their identity and location privacy. The TSE enables users to locally share service reviews on the vendors such that users receive more valuable information about the services not only from vendors but also from their trusted social friends. We study the three research topics from both theoretic and practical aspects. Specifically, we introduce the system model, review the related works, and present the solutions. We further provide the security analysis and the performance evaluation based on real-trace simulations. Lastly, we summarize our works followed by introducing the future research directions.

Waterloo, ON, Canada   Xiaohui Liang
Waterloo, ON, Canada   Rongxing Lu
Oshawa, ON, Canada   Xiaodong Lin
Waterloo, ON, Canada   Xuemin (Sherman) Shen

Contents

Chapter 1
Overview

1.1 Mobile Social Network

Social networking makes digital communication technologies sharpening tools for extending our social circle. It has already become an important integral part of our daily lives, enabling us to contact our friends and families. In the meantime, fueled by the pervasive adoption of smartphones, we have a growing tendency to access our social networks more often by smartphones than desktop computers or laptops [1]. With smartphones, we are able to check the digital personal schedules when lying in bed; read and reply to emails in the meeting room; contact friends to have a lunch together on the way to the mall; and send photos to families in the tourist areas. In other words, with smartphones, we are more capable of creating, implementing and managing of novel and efficient mobile applications. In nowadays, the pervasive use of mobile applications and the social connections fosters a promising mobile social network (MSN) where reliable, comfortable and efficient computation and communication tools are provided to improve the quality of our work and life.

1.1.1 Mobile Social Network

Over the past decade, smartphones evolve dramatically from appearance to functionality; they are no longer the clumsy equipments with basic calling and messaging functions but nice-looking and portable "toys" with integrated sensing functions and countless mobile applications. Hardware specifications of smartphones have been dramatically improved to the level of personal computers, along with friendly interface improvements and usability enhancements. In parallel to that, the deployment of 3G and LTE networks has considerably improved the available mobile bandwidth, enabling the provisioning of content and services powered by the cloud computing infrastructure. WiFi and Bluetooth techniques are pervasively used in mobile applications to enable users to communicate with

X. Liang et al., *Security and Privacy in Mobile Social Networks*, SpringerBriefs in Computer Science, DOI 10.1007/978-1-4614-8857-6_1, © The Author(s) 2013

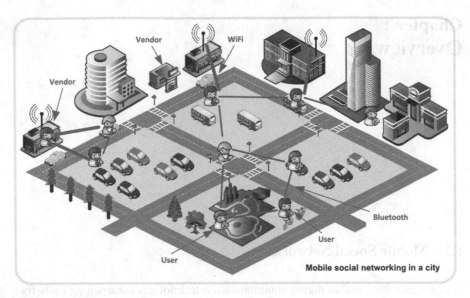

Fig. 1.1 Mobile social network

their physically-close neighbors. Due to the great potential commercial opportunity, developers and researchers design a wide range of mobile applications which can be used in different scenarios to keep up with the demand from users. As such, MSN as a pervasive communication platform to host the promising centralized/decentralized applications becomes the focus, and its research challenges and solutions are much urgent to be explored. In the following, we introduce the components of an MSN as shown in Fig. 1.1.

In the MSN, smartphones enable users to stay connected with not only the online service providers through 3G and LTE techniques, but also the physically-close peers through short-ranged wireless communication techniques, such as Bluetooth or WiFi. We consider each individual user has a unique mobility model and a personalized social behavior pattern. The reliability and efficiency of any communication protocols in the MSN are influenced by the opportunistic contacts and individual choices of users.

Vendors (a.k.a. local service providers), either mobile or static aim to provide local services to nearby users. When a vendor is mobile, it can be performed by a user who disseminates information to the encountered users in a distributed manner. When a vendor is static, it can be a local store, a restaurant, or an information center, which can be visited by the nearby users. In this case, the vendor could have more powerful and stable communication and storage devices which are placed on, in, or around their buildings.

Prior to the development of the MSN, *trusted authorities* are considered to be trusted to initialize the key materials for both vendors and users. The key materials are used to secure the local communications. Commonly-used online social systems may play the role of trusted authorities. Users often post their personal information and maintain their social connections in these systems. They are familiar with the systems and have certain trusts of the systems. The key materials can be initialized and pre-loaded into users' smartphones when users register to the system. Later on, users could authenticate others by using key materials.

1.1.2 Mobile Applications

The novelty of mobile applications is a key factor of the MSN. Recently, the number of mobile applications greatly increases. It is reported that the application store of Apple company has greatly increased the number of mobile applications available for user download from 800 in July 2008 to over 900,000 in June 2013. In the following, we divide the mobile applications into two categories, online and offline, and briefly review the popular mobile applications in each category.

1.1.2.1 Online Mobile Applications

Online mobile applications in the MSN enable users to download and upload information from and to online service providers. The typical online mobile applications are location-based services or geo-social networking services [2, 3].

Points of Interest (POI) services by Global Positioning System (GPS) devices are the earliest version of location-based services. The POIs are typically description of business storefronts. A POI could be a gas station, a hotel, a restaurant, a shopping mall, or a historical location. The information of POIs can be collected from a variety of sources, such as the mapping companies, the companies who specialize in creating business directories, or authenticated users. The data typically includes the physical address, name of the location, category of the POI, a phone number, and the geographic coordinates. The information of POIs is then stored in an online database. Users need to periodically update the POI databases of their GPS devices via the Internet connection. Then, with the geographic coordinates from the GPS devices, users can always search the nearby service information from the local POI databases. This is the first location-based service which requires users to periodically update the local databases. Note that, users' devices have to store huge amount of POI information though users may not use it at all.

Later on, the location-based services are improved along with the advancement of smartphones and networking infrastructure. Smartphones enable users to pervasively access to the Internet via WiFi or 3G/LTE networking technologies. GPS chips can be embedded in the smartphones and continuously record the locations of smartphone users. Then, the location-based service has been revolutionized in the sense that users do not need to pre-load the information of POIs before they really need it. They can report their location information to the online service providers and download local maps and nearby information of POIs in real-time. This communication pattern largely reduces the storage overhead and data management of smartphones but slightly increases the communication overhead. Supporting this pattern, the location-based services are very successful in the market. A variety of similar services have been populated in the market, by either social network giants like Facebook, Google, or specialized vendors like Foursquare, Gowalla, or Loopt.

When online service providers collect users' locations and categorize users' interests, they can not only provide location-based services but also coordinate users with similar interests and close locations to encourage more local social activities. They can further reveal the social relations under users' agreements to help them establish mutual trust and improve the communication security and efficiency. For example, the Google Latitude service enables users to visualize the locations of nearby friends on a map and launch social activities such as chatting or collaborative information filtering.

1.1.2.2 Offline Mobile Applications

While the online mobile applications require users to report their locations and download information from the online service providers in real-time, offline mobile applications can be pervasively adopted without this requirement. The offline mobile applications are more easily and likely to be launched anytime anywhere.

An interesting offline mobile application is nearby information search, where two smartphone users communicate with each other directly. This application can be very useful in many scenarios. For example, PeopleNet [4] helps a user consult her nearby friends, who in turn will ask their friends, and so on, until the information is found. Navigating for information via physically-close friends could be better than other alternative methods using the Internet because the information are more personalized and more quickly updated. The Internet applications often categorize information according to locations and communities, which is neither real-time nor easy to search. For example, the best way to know the good pizza place is to ask someone who just has eaten the pizza in the nearby store and has similar taste with the information requestor. Such information may contain trustable and friendly details in the service recommendation [5].

Carpool and ride sharing are promising solutions to the problems of urban growth and heavy traffic with more cars [6]. Besides, the increasing use of vehicle sharing and public transport is a simple, more effective way to reduce emissions as compared with the approach of producing slightly more energy efficient vehicles at a high cost. Carpooling also has a strong impact on the mode of commuting at the workplace level by only offering information, incentives and a partner matching program. Financial incentives such as reduced parking costs or fuel subsidies can drive that share higher [7]. However, in practise, users may not be willing to publish their privacy-sensitive destinations, which makes the carpool information very limited. In an offline mobile application, the direct communications between

Table 1.1 Comparison of network characteristics

	MANET	SenN	DTN	VANET	MSN
Node	◯	Sensors	◯	Vehicles	Human
Node mobility	Random	Static	◯	On-road	Human mobility
Node connectivity	◯	Good	Poor	◯	◯
Network infrastructure	No	Sink	No	RSU	No
Typical research issue	Routing	Coverage	Routing	Application	Application
Security	◯	Sensitive	◯	◯	Highly-sensitive

two smartphone users can help them share the destinations in real-time and establish
the trust relations in a distributed manner such that the chance to obtain a ride
sharing largely increases.

1.2 Characteristics of MSN

The MSN has its unique characteristics different from other traditional networks.
The following Table 1.1 enumerates the typical characteristics of mobile ad
hoc networks (MANET) [8–10], sensor networks (SenN) [11–13], delay tolerant
networks (DTN) [14–17], vehicular ad hoc networks (VANET) [17–21], and mobile
social network (MSN) [4, 5, 22–29]. ◯ means the network could have multiple
characteristics in that aspect. It is summarized that the key component of the MSN
is human. As such, the network topology relies on the human mobility. Many
research works have adopted either synthetic mobility model [30, 31] or trace-based
mobility model [32–34] to evaluate the performance of their protocols. Besides,
the functionalities of smartphones are set according to human social preferences.
In addition, smartphone users require more security and privacy protections because
most mobile applications are tightly related to their daily lives.

1.2.1 Multiple Communication Domains

Different with other networks, the MSN are considered to have multiple
communication domains where each domain has many promising mobile
applications. As shown in Fig. 1.2, depending on the applications, we consider
two-user communication domain, user-chain communication domain, and user-star
communication domain.

Two-user communication domain: Two users run a two-user communication protocol in order to obtain the personal information from each other. For example, profile
matching is a typical application in this domain where the participating users want
to know the interests of each other.

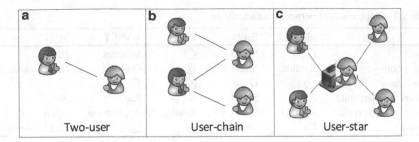

Fig. 1.2 Communication domains in MSN

User-chain communication domain: Multiple users are connected in a chain structure. They aim to cooperatively forward the data from the start user of the chain to
the end user of the chain. Due to the human mobility, the construction of chains is
unstable and mostly depends on every user's next-hop selection.

User-star communication domain: Multiple users are connected in a star topology
where the central user receives and sends the data from and to multiple users nearby.
In the service-oriented communications, the central user could be a vendor who
needs to disseminate its service information to the nearby users and receive the
service reviews from the nearby users.

1.2.2   Social Behavior

User social behavior is the foremost characteristic in the design of communication
protocols of the MSN. In this section, we will review some social theories that will
be used in later sections.

Social works [35–37] indicate that in a fully autonomous system users behave
independently based on the rational calculation of expediency. The decision on how
to act in social interactions is viewed as either primarily economic and motivated
by self-interest, or non-economic and motivated by collective interest and moral
obligation. Different norms govern users' behavior in economic and non-economic
spheres of activity, and appropriate behavior varies according to the context and
the nature of the goods being exchanged. These norms are not just learned, but
are incorporated by social users into their personalities. In reality, if users violate
a deeply internalized norm, they would feel guilty to certain extent regardless
of whether or not anyone else knew of their actions, and would likely "punish"
themselves in some manner. This is known as social morality.

Social study [38] also indicates that individuals who experience feeling of guilt
(compared to individuals who feel no guilt) after pursuing a non-cooperative strategy in the first round of play, display higher levels of cooperation in the subsequent
round of play. Experimental results demonstrate that non-cooperative individuals
who experience a certain level of guilt in a social bargaining game may use this

feeling state as "information" about the future costs of pursuing a non-cooperative strategy. Their findings that the guilt manipulation interacted with social motives (e.g., guilt tended to have its intense effect on non-cooperative individuals) also suggest that these results do not occur merely because of pre-existing individual differences in the tendency to cooperate or defect. Instead, it would appear that guilt actually provokes non-cooperative individuals to depart from their "typical" strategy of non-cooperative behavior.

The MSN users are often considered as rational and selfish entities which aim to maximize their own utility [16, 39]. In the MSN, especially when privacy enhancing technique is adopted, non-cooperative behavior is hardly overcome because users act anonymously. According to the social theories, users will always choose the behavior considering both selfishness and morality. In the design of data forwarding protocols (Chap. 3), selfishness and morality will be considered interactively. A game theoretic approach [25, 40] will be adopted to calculate the user utility of communications.

1.2.3   Social Graph

Social graph theory plays an important role in the cooperative communications of the MSN. Because user mobility is highly dynamic and unpredictable, the traditional data forwarding and dissemination protocols do not work efficiently in the MSN. Many research works [41, 42] indicate that the social relations and behaviors among the MSN users are built in a long-term fashion, which could be used to improve the communication efficiency. Specifically, from social graph theory, social community and centrality can be directly applied. For example, users in reality are often grouped based on their common interests. Sociologists found that users are more likely to interact with other same-community members [43, 44]. Thus, many research works [45–47] consider that users may encounter with another in the same community at a high probability and propose to utilize this social characteristic to better select the relays for data forwarding. Many research works [48–50] aim to identify communities from the contacts in real traces [32, 33]. Besides, centrality is another important metric in the social graph to evaluate the capability of connecting other users. To build data forwarding protocols based on this feature also improves the delivery delay [51]. In the design of anonymity enhancing techniques (Chap. 2), the social community concept will be utilized.

1.2.4   Security and Privacy

Security-related research in MSN is very critical to the success of mobile applications. Since mobile applications are often run in smartphones, the traditional security mechanisms cannot be effective due to the limited battery of smartphones.

In addition, when mobile applications are pervasively adopted in a distributed environment, security attackers can easily approach to the cyber-physical space of the target users and launch various attacks, such as forgery attacks and sybil attacks.

Forgery attacks are typical attacks in a distributed environment. Since users know little about others, the forgery attacks can easily happen. As a defense mechanism, cryptographic signature scheme [52] can be used to resist such attacks. For example, group signature [53,54] can prevent non-group members from forging a signature of group members. However, in the profile matching protocol [22, 24, 28, 55], forgery attacks on the profile matching are hardly resisted. In other words, users are able to arbitrarily choose forged profiles while other users cannot detect such behavior. Research works [24, 28] have to consider the honest-but-curious model where users honestly follow the protocol but act curiously to guess others' profiles. The work [22] requires users to make commitments about their profiles at the beginning such that they cannot change the profiles later on. However, in these works, users are always able to forge profiles without being detected. In [55], signatures of every transactions are recorded by users, and thus the forgery attacks can be caught with the help from an offline TA. However, this consumes extra communication and computation overhead.

Sybil attacks are notorious attacks in a distributed system, and they are very hard to prevent particularly when privacy is required. Such attacks subvert the system by creating a large number of pseudonymous entities, using them to gain a disproportionately large influence [56–59]. Research efforts on resisting sybil attacks have two directions. One is to study the characteristics of sybil behavior and detect the sybil accounts before the attacks [59]. The other one is to trace the sybil attacks by using cryptographic mechanisms [60].

Privacy preservation is another significant research issue in the MSN. Since personalized information are often required to be disclosed in the MSN, violating the privacy of a target user becomes much easier as indicated by many research works [23,61–66]. Research efforts [23,61,62] have been put on identification and privacy concerns in social networking sites. Gross and Acquisti [61] argued that users are putting themselves at risk both offline (e.g., stalking) and online (e.g., identity theft) based on a behavior analysis of more than 4,000 students who have joined a popular social networking site. Stutzman [62] presented a quantitative analysis of identity information disclosure in social network communities and subjective opinions from students regarding identity protection and information disclosure. When the social networking platforms are extended into the mobile environment, users require more extensive privacy-preservation because they are unfamiliar with the neighbors in close vicinity who may eavesdrop, store, and correlate their personal information at different time periods and locations. Once the personal information is correlated to the location information, the behavior of users will be completely disclosed to the public. Chen and Rahman [23] surveyed various mobile Social Networking Applications (SNAs), such as, neighborhood exploring applications, mobile-specific SNAs, and content-sharing applications, all of which provide no feedback or control mechanisms to users and may cause inappropriate location and identity information disclosure. To overcome the privacy violation in MSN, many privacy enhancing techniques have been adopted into the MSN applications [17, 22–25, 27, 28, 67, 68].

1.3 Research Topics in Mobile Social Network

We notice that users currently care much more on the functionalities of mobile applications rather than information security and privacy protections, which makes them vulnerable to the cyber-physical attacks. In this book, we will study three research topics that are related to the security and privacy in the MSN. Specifically, we will design novel communication protocols, and show that the MSN characteristics can be applied to improve the security, efficiency, and reliability of the communication protocols.

Privacy-Preserving Profile Matching (PPM): First, we study the profile matching problem. Profile matching, as the initial step of user-to-user communication, enables users to find and socialize with others who have similar interests or backgrounds. It normally requires two users to exchange their profile details so that each of them is able to check if two profiles have any similarities. However, for the constant and personalized profiles, people have growing privacy concerns in sharing it with the nearby strangers who may eavesdrop, store, and correlate their profiles at different time periods and locations. Once the profiles are identified or correlated to the location information, the behavior of users will be completely disclosed to the public. In order to preserve privacy, the privacy-preserving profile matching has been proposed: it enables two users to compare their profiles and obtain the comparison results without disclosing the profiles to each other. Many research efforts on the privacy preserving profile matching [22, 24, 27, 28, 67] have both users satisfy each other's requirement while eliminating the unnecessary information disclosure. In this book, we review some related protocols and study the user anonymity issues. We find out that the existing works do not consider user anonymity requirement at all, and the defined privacy levels cannot be related to the user-specific anonymity requirement. Thus, we take user anonymity requirement as a design goal of the profile matching protocols. We introduce an anonymity-enhancing technique for users such that users can be self-aware of the anonymity risk level and take appropriate actions to maintain the k-anonymity level. We also introduce a fully anonymous profile matching protocol which enable users to share messages and not disclose profile information at all [55].

Cooperative Data Forwarding: Second, we study the data forwarding strategy in the MSN. The MSN is considered as an opportunistic network which does not have the stable user-to-user connections due to the user mobility. As such, the effectiveness of cooperative data forwarding will largely rely on the opportunistic contacts among users. However, other than traditional opportunistic networks, the design of data forwarding strategies in the MSN must consider additional human factors, i.e., cooperative incentive and privacy preservation. In the MSN, the incentive to act cooperatively from a user's aspect includes multiple factors. Users could be cooperative because of direct friendships. In this case, it is inevitable to disclose the personal information to the encountered users which could possibly violate privacy. If users do not share personal information to others for privacy preservation, no one would act cooperatively because they have no any benefits from the cooperation. Therefore, the cooperation could be severely interrupted or even disabled when privacy preservation is applied. Many research works [40, 69–71] studied the cooperation incentives in the data forwarding and data dissemination protocols of the MSN. In this book, we will study the cooperation incentive of users following traditional social theory, and present a morality-driven data forwarding protocol with privacy preservation [25]. We will consider multiple factors in the utility of communication, such as forwarding capability, forwarding costs, and morality factor. Based on the game theoretic analysis, users always maximize their own utility and decide to forward their packets with certain probability. We are able to show that the cooperation and privacy preservation, two conflicting goals, can be achieved at the same time among a group of users with social morality.

Trustworthy Service Evaluation (TSE): Third, we study how to establish trust in the distributed MSN. In the MSN, users are often lack of trust toward others because they may not recognize the communicating opponent. Trustworthy service evaluation (TSE) system helps establish trust relationship between users and local service providers (vendors). It enables vendors to receive user feedback, known as service reviews or simply reviews, such as compliments and complaints about their services or products. By using the TSE, the vendors learn the service experiences of the users and are able to improve their service strategy in time. In addition, the collected reviews can be made available to the public, which enhances service advertising and assists the users in making wise service selections. Popular TSE can be found in web-based social networks such as Facebook, online stores like eBay, and third-party geo-social applications such as FourSquare. The TSE is often maintained by a third trusted authority who provides a platform for millions of people to interact with their friends and exchange authentic reviews. These solutions are important marketing tools for vendors who target the global market. To move the TSE into the MSN context is not easy due to the lack of third trusted authorities. In the MSN, vendors (restaurants and grocery stores) offer location-based services to local users and aim to attract the users by various advertising approaches, for example, sending e-flyers to the nearby passengers via wireless communications. Unlike the global counterparts, the interests of the vendors are in serving nearby users because most users choose services based on the comparison of the service quality and the distance advantage. Some works propose to collect the service reviews in a distributed manner and integrate the service reviews into the online location based applications. However, it is still centralized control and the review management can be complicated and cause information delay. In this book, we will introduce a distributed system where vendors maintain the TSE by themselves, and study the potential malicious attacks conducted by vendors and users. Note that, it is very challenging to restrict these malicious attacks in an untrusted and distributed environment. We will study possible review attacks and sybil attacks [60], and devise effective defensive mechanisms to resist these attacks.

1.4 Security Primitives

In this section, we will review some basic techniques.

1.4.1 K Anonymity

The k-anonymity [72] is a classic concept for evaluating anonymity. Full anonymity [73, 74] of communication protocols implies that an adversary looking at the communication patterns should not learn anything about the origin or destination of a particular message. The k-anonymity is a weaker anonymity requirement, implying that the adversary is able to learn something about the origin or destination of a particular message, but cannot narrow down its search to a set of less than k users. In other words, the k-anonymity guarantees that in a network with N honest users, the adversary is not able to guess the sender or recipient of a particular message with probability non-negligibly greater than $1/k$, where $k \leq N$ is not necessarily related to N. In practise, the parameter k can be defined by the system or individual users. When users define k, users could have different anonymity requirements and choose the appropriate strategies in the communication protocols. Using multiple pseudonym technique is a solution. When users frequently change the pseudonyms, their transactions cannot be linked at all but the number of consumed pseudonyms becomes huge. In this case, the design goal of communication protocols is to develop multiple adaptive strategies for users such that their defined anonymity requirements can be satisfied with minimum consumed pseudonyms.

1.4.2 Multiple Pseudonym Technique

A pseudonym is a name that a person assumes for a particular purpose, which differs from his or her original or true name. Pseudonyms have no literal meanings, and they can be used to hide an individual's real identity. In a network environment, pseudonyms have been widely adopted to preserve user's identity privacy [21, 75]. An offline trusted authority (TA) is considered to initialize pseudonyms for users prior to the network deployment. The TA will assign multiple pseudonyms for each individual user. These pseudonyms cannot be linked by anyone but the TA. Each user changes their pseudonyms in the communications when needed such that their behavior cannot be linked by the different pseudonyms. When users consume all the pseudonyms, they can contact with the TA to fill up with new pseudonyms.

To avoid the forgery attacks of pseudonyms, the TA assigns an additional secret to users according to their pseudonyms. Only with the secret, a user can prove that the pseudonym is legally held. The identity-based signature can be a solution. The TA generates a private key for each pseudonym, and assigns the private key to the user.

The user can always sign on any message with the private key, and the generated signature can be verified with the pseudonym. In this way, the forgery attacks of pseudonyms can be prevented.

However, with pseudonyms, users may launch malicious attacks, such as sybil attacks. To prevent the abuse of pseudonyms, the TA needs to set a trapdoor when generating the pseudonyms such that it can trace user behavior. Generally, there are two ways to implement traceable pseudonyms:

- *Mapping function*: The TA generates k pseudonyms $\{pid_{i,1}, \cdots, pid_{i,k}\}$ for user u_i. For each pseudonym $pid_{i,j}$, the TA also generates a corresponding pseudonym secret key $psk_{i,j}$ and sends the key to u_i in a secure channel. Then, u_i is able to use $pid_{i,j}$ in the communication protocols. He can generate a signature using $psk_{i,j}$ to make others confirm that u_i is the legal holder of $pid_{i,j}$. In the meantime, the TA maintains a map from these pseudonyms to the real identity id_i of u_i. When needed, others can always report the signature to the TA who is able to track u_i's behavior.
- *Group signature*: The TA generates k pairs of pseudonym and pseudonym secret key $(pid_{i,j}, psk_{i,j})$ for $1 \le j \le k$. Different from the previous method, the TA generates the pseudonym secret keys by using the key generation algorithms from group signatures [53, 54]. In this way, user u_i has to generate the group signature in the communication protocols. Although the group signature does not reveal user's real identity, the TA with a master key can always perform the trace algorithm to retrieve the real identity of the user from the group signature.

1.4.3   Prediction Method: Autoregression

The autoregressive (AR) model is a tool for understanding and predicting a time series of data [76]. It can be used to estimate the current term z_k of the series by a linear weighted sum of previous p terms (i.e., observations) in the series. The model order p is generally less than the length of the series. Formally, $\mathsf{AR}(p)$ is defined as

$$z_k = c + \sum_{i=1}^{p} \phi_o z_{k-i} + \epsilon_k, \qquad (1.1)$$

where c is a constant standing for the mean of the series, ϕ_i autoregression coefficients, and ϵ_k the zero-mean Gaussian white noise error term. For simplicity, c is often omitted.

The derivation of $\mathsf{AR}(p)$ involves determining the coefficients ϕ_i for $i \in [1 \cdots p]$ that give a good prediction. The model can be updated continuously as new samples arrive so as to ensure accuracy, or it may be recomputed when the prediction error, i.e., the difference between the predicted value and the true measurement, is very large. In [77], a simplified AR model is presented and used for neighborhood prediction. This model can be updated through trivial calculus, greatly reducing the requirement on the computational power of the nodes that implement it.

1.4.4 Cryptographic Techniques

1.4.4.1 Bilinear Groups of Prime Order

Bilinear pairing is an important cryptographic primitive and has been widely adopted in many positive applications in cryptography [78, 79]. Let $\mathbb{G}$ be a cyclic additive group and $\mathbb{G}_T$ be a cyclic multiplicative group of the same prime order q. We assume that the discrete logarithm problems in both $\mathbb{G}$ and $\mathbb{G}_T$ are hard. A bilinear pairing is a mapping $e : \mathbb{G} \times \mathbb{G} \to \mathbb{G}_T$ which satisfies the following properties:

1. *Bilinearity*: For any $P, Q \in \mathbb{G}$ and $a, b \in \mathbb{Z}_q^*$, we have $e(aP, bQ) = e(P, Q)^{ab}$.
2. *Non-degeneracy*: There exists $P \in \mathbb{G}$ and $Q \in \mathbb{G}$ such that $e(P, Q) \neq 1_{\mathbb{G}_T}$.
3. *Computability*: There exists an efficient algorithm to compute $e(P, Q)$ for all $P, Q \in \mathbb{G}$.

From [78], we note that such a bilinear pairing may be realized using the modified Weil pairing associated with supersingular elliptic curve.

Definition 1.1 (Bilinear Generator). A bilinear parameter generator $\mathcal{G}en_{bili}$ is a probability algorithm that takes a security parameter κ as input and outputs a 5-tuple $(q, P, \mathbb{G}, \mathbb{G}_T, e)$, where q is a κ-bit prime number, $(\mathbb{G}, +)$ and $(\mathbb{G}_T, \times)$ are two groups with the same order q, $P \in G$ is a generator, and $e : \mathbb{G} \times \mathbb{G} \to \mathbb{G}_T$ is an admissible bilinear map.

In the following, we briefly introduce the complexity assumptions including Computational Diffie-Hellman (CDH) problem, Decisional Diffie-Hellman (DDH) problem, Bilinear Diffie-Hellman (BDH) problem, and Decisional Bilinear Diffie-Hellman (DBDH) problem.

Definition 1.2 (Computational Diffie-Hellman (CDH) Problem). The Computational Diffie-Hellman (CDH) problem in $\mathbb{G}$ is defined as follows: given $P, aP, bP \in \mathbb{G}$ for unknown $a, b \in \mathbb{Z}_q^*$, compute $abP \in \mathbb{G}$.

Definition 1.3 (CDH Assumption). We say that an algorithm $\mathcal{A}$ has advantages $\epsilon(\kappa)$ in solving the CDH problem for $\mathbb{G}$:

$$Adv_{\mathbb{G}, \mathcal{A}}(\kappa) = \Pr[\mathcal{A}(q, \mathbb{G}, e, P, aP, bP) = P^{ab}] \geq \epsilon \qquad (1.2)$$

We say that $\mathbb{G}$ satisfies the CDH assumption if for any randomized polynomial time (in κ) algorithm $\mathcal{A}$ we have that $Adv_{\mathbb{G}, \mathcal{A}}(\kappa)$ is a negligible function. When $\mathbb{G}$ satisfies the CDH assumption we say that CDH is hard in group $\mathbb{G}$.

Definition 1.4 (Decisional Diffie-Hellman (DDH) Problem). The Decisional Diffie-Hellman (DDH) problem in $\mathbb{G}$ is defined as follows: given $P, aP, bP, cP \in \mathbb{G}$ for unknown $a, b, c \in \mathbb{Z}_q^*$, decide whether $c \stackrel{?}{=} ab$. The DDH problem in $\mathbb{G}$ is easy, since we can check whether $e(aP, bP) \stackrel{?}{=} e(P, cP)$ and use the results to decide $c \stackrel{?}{=} ab$.

Definition 1.5 (Bilinear   Diffie-Hellman   (BDH)   Problem). The   Bilinear Diffie-Hellman (BDH) problem in $\mathbb{G}$ is defined as follows: given $P, aP, bP, cP \in \mathbb{G}$ for unknown $a, b, c \in \mathbb{Z}_q^*$, compute $e(P, P)^{abc} \in \mathbb{G}_T$.

Definition 1.6 (BDH Assumption). We say that an algorithm $\mathscr{A}$ has advantages $\epsilon(\kappa)$ in solving the BDH problem for $\mathscr{G}en_{bili}$ if for sufficiently large k:

$$Adv_{\mathscr{G}en_{bili},\mathscr{A}}(\kappa) = \Pr[\mathscr{A}(q, \mathbb{G}, \mathbb{G}_T, e, P, aP, bP, cP) = e(P, P)^{abc}] \geq \epsilon(\kappa) \quad (1.3)$$

We say that $\mathscr{G}en_{bili}$ satisfies the BDH assumption if for any randomized polynomial time (in κ) algorithm $\mathscr{A}$ we have that $Adv_{\mathscr{G}en_{bili},\mathscr{A}}(\kappa)$ is a negligible function. When $\mathscr{G}en_{bili}$ satisfies the BDH assumption we say that BDH is hard in groups generated by $\mathscr{G}en_{bili}$.

Definition 1.7 (Decisional Bilinear Diffie-Hellman (DBDH) Problem). The Decisional Bilinear Diffie-Hellman (DBDH) problem in $\mathbb{G}$ is defined as follows: given P, aP, bP, cP, T for unknown $a, b, c \in \mathbb{Z}_q^*$ and $T \in \mathbb{G}_T$, decide whether $T \overset{?}{=} e(P, P)^{abc}$.

Definition 1.8 (DBDH Assumption). We say that an algorithm $\mathscr{A}$ has advantages $\epsilon(\kappa)$ in solving the DBDH problem for $\mathscr{G}en_{bili}$ if for sufficiently large k, $\mathscr{A}$ distinguishes the two tuples (P, aP, bP, cP) and (P, aP, bP, abP) with advantage $\epsilon(\kappa)$, i.e.,

$$| \Pr[\mathscr{A}(q, \mathbb{G}, \mathbb{G}_T, e, P, aP, bP, cP, e(P, P)^{abc}) = 1]$$
$$- \Pr[\mathscr{A}(q, \mathbb{G}, \mathbb{G}_T, e, P, aP, bP, cP, T) = 1]| \geq \epsilon(\kappa) \quad (1.4)$$

We say that $\mathscr{G}en_{bili}$ satisfies the DBDH assumption if for any randomized polynomial time (in κ) algorithm $\mathscr{A}$, it distinguishes the two tuples with a negligible probability. When $\mathscr{G}en_{bili}$ satisfies the DBDH assumption we say that DBDH is hard in groups generated by $\mathscr{G}en_{bili}$.

1.4.4.2 Bilinear Groups of Composite Order

In the previous definition, groups $\mathbb{G}$ and $\mathbb{G}_T$ have the same prime order q. In literature [53, 54], there have been many cryptographic scheme design using the bilinear groups of composite order. Generally, these works can provide additional anonymity protection and trace capability. We briefly review its definition as follows.

Let two finite cyclic groups $\mathbb{G}$ and $\mathbb{G}_T$ having the same order n, in which the respective group operation is efficiently computable and denoted multiplicatively [53, 54]. Assume that there exists an efficiently computable function $e : \mathbb{G} \times \mathbb{G} \rightarrow \mathbb{G}_T$, called a bilinear map or pairing, with the following properties:

1. *Bilinearity*: For any $u, v \in \mathbb{G}$ and $a, b \in \mathbb{Z}_q^*$, we have $e(u^a, v^b) = e(u, v)^{ab}$.
2. *Non-degeneracy*: $\exists g \in \mathbb{G}$ such that $e(g, g)$ has order n in $\mathbb{G}_T$. In other words, $e(g, g)$ is a generator of $\mathbb{G}_T$, whereas g generates $\mathbb{G}$.

Note that, the operation in $\mathbb{G}$ is denoted as multiplication, which is just for easy presentation. The bilinear groups of composite order differ from the previous ones by changing a prime order p to a composite order $n = pq$ where $p \neq q$ are two large primes. The factorization problem of n is assumed to be computationally-infeasible. The complexity assumptions in the bilinear groups of prime order also hold in the bilinear groups of composite order. In addition, we introduce the SubGroup Decision (SGD) Problem as follows:

Definition 1.9 (SubGroup Decision (SGD) Problem). The SubGroup Decision (SGD) problem in $\mathbb{G}$ is defined as follows: given $(e, \mathbb{G}, \mathbb{G}_T, n, h)$ where the element h is randomly drawn from either $\mathbb{G}$ or subgroup $\mathbb{G}_q$, decide whether $h \in \mathbb{G}_q$ or $h \in \mathbb{G}$.

Definition 1.10 (SGD Assumption). We say that an algorithm $\mathscr{A}$ has advantages $\epsilon(\kappa)$ in solving the SGD problem for $\mathbb{G}$ and $\mathbb{G}_q$ if for sufficiently large k, $\mathscr{A}$ correctly guess either $h \in \mathbb{G}_q$ or $h \in \mathbb{G}$ with advantage $\epsilon(\kappa)$, i.e.,

$$| \Pr[\mathscr{A}(h \in \mathbb{G}_q) = 1] - \Pr[\mathscr{A}(q \in \mathbb{G}) = 1]| \geq \epsilon(\kappa) \tag{1.5}$$

We say that $\mathbb{G}$ and $\mathbb{G}_q$ satisfy the SGD assumption if for any randomized polynomial time (in κ) algorithm $\mathscr{A}$, it correctly guesses either $h \in \mathbb{G}_q$ or $h \in \mathbb{G}$ with a negligible probability. When $\mathscr{G}en_{bili}$ satisfies the SGD assumption we say that SGD is hard in groups $\mathbb{G}$ and $\mathbb{G}_q$.

1.4.4.3 Identity Based Aggregate Signature

The identity based aggregate signature (IBAS) scheme [80] consists of five algorithms, *Setup*, *Private key generation*, *Individual Signing*, *Aggregation*, and *Verification*.

- *Setup*: The Private Key Generator (PKG) uses a bilinear generator $\mathscr{G}en_{bili}$ from the previous section to generate a 5-tuple $(q, P, \mathbb{G}, \mathbb{G}_T, e)$. The PKG also picks a random $s \in \mathbb{Z}_q$ and sets $Q = sP$. It chooses a cryptographic hash functions $H_1, H_2 : \{0, 1\}^* \to \mathbb{G}_1$ and $H_3 : \{0, 1\}^* \to \mathbb{Z}_q$.
- *Private key generation*: The user u_i with identity ID_i receives from the PKG the private key $sP_{i,j}$ for $j \in \{0, 1\}$, where $P_{i,j} = H_1(ID_i, j) \in \mathbb{G}$.
- *Individual signing*: The first signer chooses a string w that it has never used before. Each subsequent signer checks that it has not used the string w chosen by the first signer. To sign m_i, the signer with identity ID_i:

  1. computes $P_w = H_2(w) \in \mathbb{G}$;
  2. computes $c_i = H_3(m_i, ID_i, w) \in \mathbb{Z}_q$;

3. generates random $r_i \in \mathbb{Z}_q$;
4. computes its signature (w, S_i', T_i'), where $S_i' = r_i P_w + s P_{i,0} + c_i s P_{i,1}$ and $T_i' = r_i P$.

- *Aggregation*: Anyone can aggregate a collection of individual signatures that use the same string w. For example, individual signatures (w, S_i', T_i') for $1 \leq i \leq n$ can be aggregated into (w, S_n, T_n), where $S_n = \sum_{i=1}^{n} S_i'$ and $T_n = \sum_{i=1}^{n} T_i'$.
- *Verification*: Let (w, S_n, T_n) be the identity-based aggregate signature where n is the number of signers. The verifier checks that:

$$e(S_n, P) = e(T_n, P_w) e(Q, \sum_{i=1}^{n} P_{i,0} + \sum_{i=1}^{n} c_i P_{i,1}), \tag{1.6}$$

where $P_{i,j} = H_1(ID_i, j)$, $P_w = H_2(w)$ and $c_i = H_3(m_i, ID_i, w)$.

1.4.4.4   Shamir Secret Sharing

The goal of a secret sharing scheme [81] is to divide a secret s into n pieces $s_1, s_2, \cdots, s_n$ in such as way that:

- knowledge of any k or more s_i pieces makes s easily computable;
- knowledge of any $k - 1$ or fewer s_i pieces leaves s completely undetermined (in the sense that all its possible values are equally likely).

Such a scheme is called a (k, n) threshold scheme.

The following scheme [81] is an example of a (k, n) secret sharing. Denote $\mathbb{G}$ as a group. Suppose that a user has a secret $s \in \mathbb{G}$.

- *Secret shares generation*: To generate n s_i, the user chooses a polynomial with a degree k, i.e., $f(x) = x^k + a_{k-1} x^{k-1} + \cdots a_1 x + a_0 (= s)$ where $a_0, \cdots, a_{k-1}$ are randomly chosen from $\mathbb{G}$. Then, the user generates $s_i = f(i)$ for $i = 1, \cdots, n$.
- *Secret recovery*: To recover the secret s, a lagrange interpolation is used to calculate s from n pieces $s_1, \cdots, s_n$ following the equation

$$s = f(0) = \sum_{i=1}^{n} \alpha_i \cdot s_i, \text{ where } \alpha_i = \prod_{j=1, j \neq i}^{n} \frac{-j}{i - j}. \tag{1.7}$$

1.4.4.5   Homomorphic Encryption

There are several existing homomorphic encryption schemes that support different operations such as addition and multiplication on ciphertexts, e.g. [82,83]. By using these schemes, a user is able to process the encrypted plaintext without knowing the secret keys. Due to this property, homomorphic encryption schemes are widely used in data aggregation and computation specifically for privacy-sensitive content [84]. We review the homomorphic encryption scheme [83].

Suppose user u_i has a public/private key pair (pk_i, sk_i) from the fully homomorphic encryption (FHE) scheme. The Encryption Enc, Decryption Dec, Addition Add, and Multiplication Mul functions must be satisfied.

- Correctness: $Dec(sk_i, Enc(pk_i, m)) = m$;
- Addition of plaintexts: $Dec(sk_i, Add(Enc(pk_i, m_1), Enc(pk_i, m_2))) = m_1 + m_2$;
- Multiplication of plaintexts: $Dec(sk_i, Mul(Enc(pk_i, m_1), Enc(pk_i, m_2))) = m_1 \cdot m_2$.

The following scheme is an example of the FHE scheme. A trusted authority runs a generator $\mathcal{G}en_{homo}$ which outputs $\langle p, q, R, R_q, R_p, \chi \rangle$ as system public parameters:

- $p < q$ are two primes s.t. $q \equiv 1 \mod 4$ and $p \gg l$;
- Rings $R = \mathbb{Z}[x]/\langle x^2 + 1 \rangle$, $R_q = R/qR = \mathbb{Z}_q[x]/\langle x^2 + 1 \rangle$;
- Message space $R_p = \mathbb{Z}_p[x]/\langle x^2 + 1 \rangle$;
- A discrete Gaussian error distribution $\chi = D_{\mathbb{Z}^n, \sigma}$ with standard deviation σ.

Suppose user u_i has a public/private key pair (pk_i, sk_i) such that $pk_i = \{a_i, b_i\}$, with $a_i = -(b_i s + pe)$, $b_i \in R_q$ and $s, e \leftarrow \chi$, and $sk_i = s$. Let $b_{i,1}$ and $b_{i,2}$ be two messages encrypted by u_i.

- Encryption $E_{pk_i}(b_{i,1})$: $c_{i,1} = (c_0, c_1) = (a_i u_t + p g_t + b_{i,1}, b_i u_t + p f_t)$, where u_t, f_t, g_t are samples from χ.
- Decryption $D_{sk_i}(c_{i,1})$: If denoting $c_{i,1} = (c_0, \cdots, c_{\alpha_1})$, $b_{i,1} = (\sum_{k=0}^{\alpha_1} c_k s^k)$ mod p.

Consider the two ciphertexts $c_{i,1} = E(b_{i,1}) = (c_0, \cdots, c_{\alpha_1})$ and $c_{i,2} = E(b_{i,2}) = (c_0', \cdots, c_{\alpha_2}')$.

- Addition: Let $\alpha = max(\alpha_1, \alpha_2)$. If $\alpha_1 < \alpha$, let $c_{\alpha_1+1} = \cdots = c_\alpha = 0$; If $\alpha_2 < \alpha$, let $c_{\alpha_2+1}' = \cdots = c_\alpha' = 0$. Thus, we have $E(b_{i,1} + b_{i,2}) = (c_0 \pm c_0', \cdots, c_\alpha \pm c_\alpha')$.
- Multiplication: Let v be a symbolic variable and compute $(\sum_{k=0}^{\alpha_1} c_k v^k) \cdot (\sum_{k=0}^{\alpha_2} c_k' v^k) = \hat{c}_{\alpha_1+\alpha_2} v^{\alpha_1+\alpha_2} + \cdots + \hat{c}_1 v + \hat{c}_0$. Thus, we have $E(b_{i,1} \times b_{i,2}) = (\hat{c}_0, \cdots, \hat{c}_{\alpha_1+\alpha_2})$.

Chapter 2
Profile Matching Protocol with Anonymity Enhancing Techniques

2.1 Introduction

In this chapter, we introduce a popular research topic, called privacy-preserving profile matching (PPM). The PPM can be very useful in an application scenario where two users both want to know something about each other but they do not want to disclose too much personal information. The PPM occurs quite frequently in our daily lives. For example, in a restaurant or a sports stadium, people like finding their friends and chatting with friendly neighbors. To initialize the communication, they may expect to know if others have similar preferences or share the similar opinions. In a mobile healthcare system, patients may be willing to share personal symptoms with others. However, they expect the listeners to have similar experiences such that they could receive comforts and suggestions. Based on the above application scenarios, we can see that the common design goal of the PPM is to help two users exchange personal information while preserving their privacy. It could serve as the initial communication step of many mobile social networking applications.

We will introduce a family of the PPM protocols. Specifically, these protocols rely on the homomorphic encryption to protect the content of user profiles from disclosure. They provide increasing levels of user anonymity (from conditional to full). Furthermore, we will study the social community concept and adopt the prediction method and the multiple-pseudonym technique to improve the user anonymity protection in the protocol. The extensive trace-based simulation results show that the protocol with the anonymity enhancing technique achieves significantly higher anonymity strength with slightly larger number of used pseudonyms than the protocol without the technique.

The remainder of this chapter is organized as follows: in Sect. 2.2, we present the network model and the design goal. Then, we introduce three protocols in Sect. 2.3. We introduce the anonymity enhancing techniques in Sect. 2.4 and provide the simulation-based performance evaluations of the protocols and techniques in Sect. 2.5. Lastly, we review the related work and draw our summary respectively in Sects. 2.6 and 2.7.

X. Liang et al., *Security and Privacy in Mobile Social Networks*, SpringerBriefs in Computer Science, DOI 10.1007/978-1-4614-8857-6_2, © The Author(s) 2013

2.2 Network Model and Design Goal

2.2.1 Network Model

We consider a homogenous MSN composed of a set $\mathcal{V} = \{u_1, \cdots, u_N\}$ of mobile users with the network size $|\mathcal{V}| = N$. Users have equal wireless communication range, and the communication is bi-directional. Each user obtains multiple pseudonyms from a trusted authority (TA) (Sect. 1.4.2) and uses these pseudonyms instead of their real identities to preserve their privacy.

Profile: Each user has a profile spanning w attributes $W = \{a_1, \cdots, a_w\}$. These w attributes include every aspects of a user. Each profile is represented by a w-dimension vector where each dimension has an integer value between 1 and l. The profile of user u_i is denoted by $p_i = (v_{i,1}, \cdots, v_{i,w})$ where $v_{i,h} \in \mathbb{Z}$ and $1 \le v_{i,h} \le l$ for $1 \le h \le w$.

Matching operation: A matching operation between two profiles $p_i = (v_{i,1}, \cdots, v_{i,w})$ and $p_j = (v_{j,1}, \cdots, v_{j,w})$ can be

- **Inner product** $f_{dot}(p_i, p_j) = f_{dot}(p_i, p_j) = \sum_{t=1}^{w} v_{i,t} \cdot v_{j,t}$.
- **Manhattan distance** $f_{man}(p_i, p_j, \alpha) = f_{man}(p_i, p_j) = (\sum_{t=1}^{w} |v_{i,t} - v_{j,t}|^\alpha)^{\frac{1}{\alpha}}$.
- **Max distance** $f_{max}(p_i, p_j) = max\{|v_{i,1} - v_{j,1}|, \cdots, |v_{i,w} - v_{j,w}|\}$.
- **Comparison-based**

$$f_{cmp}(p_i, p_j, x) = \begin{cases} -1, v_{i,x} < v_{j,x} \\ 0, v_{i,x} = v_{j,x} \\ 1, v_{i,x} > v_{j,x} \end{cases} \quad (2.1)$$

- **Predicate-based**

$$f_{cmp}(p_i, p_j, \Pi) = \begin{cases} 1, (p_i, p_j) \in \Pi \\ -1, (p_i, p_j) \notin \Pi \end{cases} \quad (2.2)$$

Note, $\Pi = $ "$\bar{t}$ of $\{(v_{i,x}, opt, v_{j,x}) | a_x \in A\}$" is a predicate where $A \subseteq W$ and the comparison operator opt is either $>$ or $<$ and $\bar{t} \le |A|$. $f_{cmp}(p_i, p_j, \Pi) = 1$ if (p_i, p_j) satisfies at least $\bar{t}$ equations; $f_{cmp}(p_i, p_j, \Pi) = -1$ otherwise.

2.2.2 Design Goal

The objective of a privacy-preserving profile matching protocol is to enable two users to compare their profiles while not disclosing the profiles to each other. The matching operations in previous section can be easily done if (p_i, p_j) can be obtained by any single user. However, users will not disclose their profiles p_i and

p_j to others, and no trusted authority exists. This makes the implementation of the matching operations very difficult. Recent secure multi-party computation (MPC) is developed to enable users to jointly compute a function over their inputs, while at the same time keeping these inputs private. The two-party computation can be used for profile matching. In addition, many recent works [22, 28] realize that the explicit matching result may reveal the uniqueness of profiles, and extend their protocols to show fuzzy matching results instead of explicit results, i.e., users are able to know if the matching result is larger or smaller than a pre-defined threshold. In the following, we review three kinds of matching results.

- Explicit matching result: the matching result $f_*()$ is directly disclosed to the participants u_i and/or u_j.
- Fuzzy matching result: the relationship between the matching result $f_*()$ and a predefined threshold T (i.e., if $f_*() > T$) is disclosed to the participants u_i and/or u_j.
- Implicit matching result: a message $F_i(f_*())$ is implicitly disclosed to the participant u_j where $F_i()$ is a secret mapping function defined by u_i and u_i is unaware of both $f_*()$ and $F_i(f_*())$.

Note that, the amount of disclosed information is reduced from explicit to fuzzy to implicit matching results. In practise, users expect to know how much profile information disclosed in each profile matching protocol. They do not want to over-disclose the profile information such that their behavior will not be tracked. Especially, when the disclosed information can be linked in multiple runs, the behavior of an individual user will be more easily to track. We consider that users apply the multiple pseudonym technique (Sect. 1.4.2). Then, the problem will be how to change the pseudonyms such that the leaked information is not be enough to uniquely identify a user and how to reduce the number of used pseudonyms.

The multiple pseudonym technique gives users the capability to self-protect their private information. Therefore, instead of minimizing the disclosed information in each protocol run, we will focus on how to protect user anonymity in the multiple protocol runs. To address this point, we define anonymity levels for the profile matching protocols as the probability of correctly guessing the profile of the user from the profile samples. If a user does not change the pseudonym, its anonymity continuingly decreases as the protocol runs because the disclosed information will help narrow down the possibilities of the profile. If a user changes the pseudonym, its anonymity is changed to the highest level. Considering a user could possibly have v profiles from an attacker's point of view, we define three types of anonymity levels for profile matching protocols, i.e., non-anonymity, conditional anonymity, and full anonymity.

Definition 2.1 (Non-anonymity). A profile matching protocol provides *non-anonymity* if after executing multiple runs of the protocol with any user, the probability of correctly guessing the profile of the user is equal to 1.

Definition 2.2 (Conditional Anonymity). A profile matching protocol achieves *conditional anonymity* if after executing multiple runs of the protocol with some user, the probability of correctly guessing the profile of the user is larger than $\frac{1}{v}$.

Definition 2.3 (Full Anonymity). A profile matching protocol achieves *full anonymity* if after executing multiple runs of the protocol with any user, the probability of correctly guessing the profile of the user is always $\frac{1}{v}$.

Compared with the previous definitions on matching results, user anonymity is a user-centric privacy preservation method for the profile matching protocols, which has the following advantages.

1. The user anonymity is defined in multiple runs while the level privacy level only relates to the output of a single protocol run.
2. The user anonymity is a user-centric requirement while the previous privacy level is the protocol-specific requirement. The user anonymity could provide a practical solution for users to self-protect their privacy.

In most recent works [22, 28], the explicit and fuzzy matching results could be possibly used to track user behavior. Thus, these works provide non-anonymity and conditional anonymity. In comparison, the implicit matching results do not reveal the profile information and the protocol that discloses implicit matching results can achieve full anonymity. In the following, we introduce three comparison-based profile matching protocols: an explicit Comparison-based Profile Matching protocol (eCPM), an implicit Comparison-based Profile Matching protocol (iCPM), and an implicit Predicate-based Profile Matching protocol (iPPM) [55]. The eCPM achieves conditional anonymity while the eCPM and the iCPM achieve full anonymity.

2.3 PPM Solutions

We first use an example to describe the profile matching with the comparison operation. Consider two CIA agents have two different rank levels in the CIA system, A with a low priority l_A and B with a high priority l_B. They know each other as a CIA agent. However, they do not want to reveal their priority levels to each other. B wants to share some messages to A. The messages are not related to user profile, and they are divided into multiple categories, e.g., the messages related to different regions (New York or Beijing) in different years (2011 or 2012). B shares one message of a specified category T at a time. The category T is chosen by A, but the choice is unknown to B. For each category, B prepares two self-defined messages, e.g., a message m_1 for the CIA agent at a lower level and another message m_2 for the agent at a higher level. Because $l_A < l_B$, A eventually obtains m_1. In the meantime, B does not know which message A receives. The above profile matching offers both A and B the highest anonymity since neither the comparison result between l_A and l_B is disclosed to A or B nor the category T of A's interest is

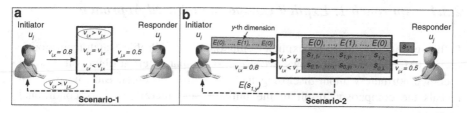

Fig. 2.1 Scenarios

disclosed to B. In the following, we refer to A as the initiator u_i, B as the responder u_j, the attribute used in the comparison (i.e., priority level) as a_x, and the category T of A's interest as T_y. The attribute values of u_i and u_j on the attribute a_x are denoted by $v_{i,x}$ and $v_{j,x}$, respectively. We first formally describe two scenarios from the above examples: (a) attribute value $v_{i,x}$ and attribute value $v_{j,x}$ will not be disclosed to u_j and u_i, respectively. The initiator obtains the comparison result at the end of the protocol. (b) $v_{i,x}$ and $v_{j,x}$ will not be disclosed to u_j and u_i, respectively. In addition, category T_y will not be disclosed to u_j, and the comparison result will not be disclosed to any of u_i and u_j. The initiator obtains either $s_{1,y}$ or $s_{0,y}$ depending on the comparison result between $v_{i,x}$ and $v_{j,x}$.

Scenario-1: The initiator wants to know the comparison result, i.e., whether it has a value larger, equal, or smaller than the responder on a specified attribute. For example, as shown in Fig. 2.1a, the initiator u_i expects to know if $v_{i,x} > v_{j,x}$, $v_{i,x} = v_{j,x}$, or $v_{i,x} < v_{j,x}$.

Scenario-2: The initiator expects that the responder shares one message related to the category of its interest, which is however kept unknown to the responder. In the meantime, the responder wants to share with the initiator one message which is determined by the comparison result of their attribute values. For example, as shown in Fig. 2.1b, both u_i and u_j know that a_x is used in the comparison and the categories of messages are $T_1, \cdots, T_\lambda$. The initiator u_i first generates a $(0, 1)$-vector where the y-th dimension value is 1 and other dimension values are 0. Then, u_i encrypts the vector with its own public key and sends the ciphertexts $(E(0), \cdots, E(1), \cdots, E(0))$ to the responder u_j. The ciphertexts imply u_i's interested category T_y, but u_j is unable to know T_y since $E(0)$ and $E(1)$ are non-distinguishable without a decryption key. u_i also provides its attribute value $v_{i,x}$ in an encrypted form so that u_j is unable to obtain $v_{i,x}$. On the other hand, u_j prepares λ pairs of messages, each pair $(s_{1,h}, s_{0,h})$ relating to one category $T_h(1 \le h \le \lambda)$. u_j executes a calculation over the ciphertexts and sends the result to u_i. Finally, u_i obtains $E(s_{1,y})$ if $v_{i,x} > v_{j,x}$ or $E(s_{0,y})$ if $v_{i,x} < v_{j,x}$, and obtains $s_{1,y}$ or $s_{0,y}$ by the decryption.

2.3.1 Approach 1: Explicit Comparison-Based Approach

In this section, we present the explicit Comparison-based Profile Matching protocol, i.e., eCPM. This protocol allows two users to compare their attribute values on a specified attribute without disclosing the values to each other. But, the protocol reveals the comparison result to the initiator, and therefore offers conditional anonymity.

Bootstrapping: The protocol has a fundamental bootstrapping phase, where the TA generates all system parameters, user pseudonyms, and keying materials. Specifically, the TA runs $\mathcal{G}$ to generate $\langle p, q, R, R_q, R_p, \chi \rangle$ for initiating the homomorphic encryption (see Sect. 1.4.4.5). The TA generates a pair of public and private keys (pk_{TA}, sk_{TA}) for itself. The public key pk_{TA} is open to all users; the private key sk_{TA} is a secret which will be used to issue certificates for user pseudonyms and keying materials, as shown below.

The TA generates disjoint sets of pseudonyms (pid_i) and disjoint sets of homomorphic public keys (pk_i) for users (u_i). For every pid_i and pk_i of u_i, the TA generates the corresponding secret keys psk_i and sk_i. In correspondence to each pseudonym pid_i, it assigns a certificate $cert_{pid_i}$ to u_i, which can be used to confirm the validity of pid_i. Generally, the TA uses sk_{TA} to generate a signature on pid_i and pk_i. The TA outputs $cert_{pid_i}$ as a tuple $(pk_i, Sign_{sk_{TA}}(pid_i, pk_i))$. The homomorphic secret key sk_i is delivered to u_i together with psk_i; pk_i is tied to pid_i and varies as the change of pseudonyms.

Protocol Steps: Consider user u_i with a neighboring user u_j. Denote by pid_i the current pseudonym of u_i and by pid_j that of u_j. Recall that a_x is an attribute, $v_{i,x}$ and $v_{j,x}$ the values of u_i and u_j on a_x, and l the largest attribute value. Suppose that u_i as an initiator starts profile matching on a_x with a responder u_j. Let psk_i and psk_j be the secret keys corresponding to pid_i and pid_j, respectively. The protocol is executed as follows.

Step 1. u_i calculates $d_i = E_{pk_i}(v_{i,x})$, and sends the following 5-tuple to u_j.

$$(pid_i, cert_{pid_i}, a_x, d_i, Sign_{psk_i}(a_x, d_i))$$

Step 2. After receiving the 5-tuple, u_j opens the certificate $cert_{pid_i}$ and obtains the homomorphic public key pk_i and a signature. It checks $cert_{pid_i}$ to verify that (pk_i, pid_i) are generated by the TA, and it checks the signature to validate (a_x, d_i). If any check is failed, u_j stops; otherwise, u_j proceeds as follows. It uses pk_i to encrypt its own attribute value $v_{j,x}$, i.e., $d_j = E_{pk_i}(v_{j,x})$; it chooses a random value $\varphi \in \mathbb{Z}_p$ such that $1 \leq \varphi < \lfloor p/(2l) \rfloor$ and $m|\varphi$ for any integer $m \in [1, l-1]$ (φ can be chosen dependent on u_j's anonymity requirement). By the homomorphic property, it calculates $E_{pk_i}(v_{i,x} - v_{j,x})$ and $d'_j = E_{pk_i}(\varphi(v_{i,x} - v_{j,x}))$; it finally sends a 5-tuple $(pid_j, cert_{pid_j}, a_x, d'_j, Sign_{psk_j}(a_x, d'_j))$ to u_i.

Step 3. After receiving the 5-tuple, u_i opens the certificate $cert_{pid_j}$ and checks the signature to make sure the validity of pid_j and (a_x, d'_j). If the check is successful, u_i uses sk_i to decrypt d'_j and obtains the comparison result $c = \varphi(v_{i,x} - v_{j,x})$. u_i knows $v_{i,x} > v_{j,x}$ if $0 < c \leq \frac{p-1}{2}$, $v_{i,x} = v_{j,x}$ if $c = 0$, or $v_{i,x} < v_{j,x}$ otherwise.

Effectiveness Discussion: The effectiveness of the eCPM is guaranteed by the following theorems.

Theorem 2.1 (Correctness). *In the eCPM, the initiator u_i is able to obtain the correct comparison result with the responder u_j on a specified attribute a_x.*

Proof. Recall $p \gg l$ and $1 \leq \varphi < \lfloor p/(2l) \rfloor$. As $1 \leq v_{i,x}, v_{j,x} \leq l$, we have $-l < v_{i,x} - v_{j,x} < l$. If $v_{i,x} > v_{j,x}$, we have $0 < \varphi(v_{i,x} - v_{j,x}) < \lfloor p/(2l) \rfloor \times l \leq p/2$. Because p is a prime and $\varphi(v_{i,x} - v_{j,x})$ is an integer, we have $0 < \varphi(v_{i,x} - v_{j,x}) \leq (p-1)/2$. In case of $v_{i,x} < v_{j,x}$, we may similarly derive $(p+1)/2 \leq \varphi(v_{i,x} - v_{j,x}) < p$. Thus, by comparing $\varphi(v_{i,x} - v_{j,x})$ with 0, $(p-1)/2$ and $(p+1)/2$, u_i is able to know whether $v_{i,x} > v_{j,x}$, $v_{i,x} = v_{j,x}$, or $v_{i,x} < v_{j,x}$.   □

Theorem 2.2 (Anonymity). *The eCPM does not disclose the attribute values of participating users.*

Proof. The initiator u_i who starts the protocol for attribute a_x encrypts its attribute value $v_{i,x}$ using its homomorphic public key pk_i. Thus, the responder u_j is unable to know any information about $v_{i,x}$. On the other side, the responder u_j does not transmit its attribute value $v_{j,x}$, but returns $\varphi(v_{i,x} - v_{j,x})$ to u_i, where φ is a random factor added for anonymity. Since $m | \varphi$ for $1 \leq m \leq l-1$, we have $m | (\varphi(v_{i,x} - v_{j,x}))$. Thus, $(v_{i,x} - v_{j,x})$ can be any value between $-(l-1)$ and $l-1$ from u_i's view, and the exact value of $v_{j,x}$ is thus protected.   □

Theorem 2.3 (Non-forgeability). *The eCPM discourages profile forgery attack at the cost of involving the TA for signature verification and data decryption.*

Proof. Consider two users u_i and u_j running the eCPM with each other on attribute a_x. Their public keys pk_i and pk_j used for homomorphic encryption are generated by the TA, and the TA has full knowledge of the corresponding private keys sk_i and sk_j. In addition, their attribute values are generated by the TA and recorded in the TA's local repository, and the TA can retrieve any attribute value of users (e.g. $v_{i,x}$ or $v_{j,x}$) anytime when necessary. After the two users finish the protocol, u_i will have $Sign_{psk_j}(d'_j)$, and u_j will have $Sign_{psk_i}(d_i)$. If $u_i(u_j)$ uses the forged profile in the protocol, $u_j(u_i)$ can cooperate with the TA to trace such malicious attack. Specifically, $u_j(u_i)$ can send $Sign_{psk_i}(d_i)$ ($Sign_{psk_j}(d'_j)$) to the TA. the TA will be able to check if the signatures are valid and the encrypted values are consistent with $v_{i,x}$ and $v_{j,x}$. Thus, any profile forgery attack can be detected with the help from the TA, and such attacks will be discouraged.   □

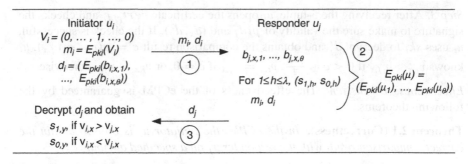

Fig. 2.2 The iCPM flow

2.3.2   Approach 2: Implicit Comparison-Based Approach

In this section, we introduce the implicit Comparison-based Profile Matching (iCPM) by adopting the oblivious transfer cryptographic technique [85]. We consider users have distinct values for any given attribute. As shown in Fig. 2.2, the iCPM consists of three main steps. In the first step, u_i chooses an interested category T_y by setting y-th element to 1 and other elements to 0 in a λ-length vector V_i. u_i then encrypt the vector by using the homomorphic encryption and sends the encrypted vector to u_j. Thus, u_j is unable to know T_y but still can process on the ciphertext. In the second step, u_j computes the ciphertexts with input of self-defined messages $(s_{1,h}, s_{0,h})$ for $1 \le h \le \lambda$, two encrypted vectors (m_i, d_i), and its own attribute value $v_{j,x}$. In the last step, u_i decrypts the ciphertext and obtain $s_{1,y}$ if $v_{i,x} > v_{j,x}$ or $s_{0,y}$ if $v_{i,x} < v_{j,x}$.

Protocol Steps: In the iCPM, the responder u_j prepares λ pairs of messages $(s_{0,h}, s_{1,h})$ for category T_h $(1 \le h \le \lambda)$ where $s_{0,h}, s_{1,h} \in \mathbb{Z}_p$ and $s_{0,h}, s_{1,h} \le (p-1)/2$. These messages are not related to u_j's profile. The initiator u_i first decides which category T_y it wants to receive messages related to. But u_i does not disclose T_y to u_j. Then, the responder u_j shares either $s_{0,y}$ or $s_{1,y}$ to u_i without knowing which one will be received by u_i. When the protocol finishes, u_i receives one of $s_{0,y}$ and $s_{1,y}$ with no clue about the comparison result. We elaborate the protocol steps below.

Step 1. u_i generates a vector $V_i = (v_1, \cdots, v_\lambda)$, where $v_y = 1$ and $v_h = 0$ for $1 \le h \le \lambda$ and $h \ne y$. This vector implies that u_i is interested in the category T_y. u_i sets $m_i = E_{pk_i}(V_i) = (E_{pk_i}(v_1), \cdots, E_{pk_i}(v_\lambda))$. It converts $v_{i,x}$ to binary bits $\langle b_{i,x,1}, \cdots, b_{i,x,\theta}\rangle$, where $\theta = \lceil \log l \rceil$, and sets $d_i = (E_{pk_i}(b_{i,x,1}), \cdots, E_{pk_i}(b_{i,x,\theta}))$. It sends u_j a 6-tuple $(pid_i, cert_{pid_i}, a_x, d_i, m_i, Sign_{psk_i}(a_x, d_i, m_i))$.

Step 2. After receiving the 6-tuple, u_j checks if $(pid_i, cert_{pid_i})$ are generated by the TA and the signature is generated by u_i. If both checks are successful, it knows that (a_x, d_i, m_i) is valid. u_j proceeds as follows:

1. Convert $v_{j,x}$ to binary bits $\langle b_{j,x,1}, \cdots, b_{j,x,\theta}\rangle$ and compute $E_{pk_i}(b_{j,x,t})$ for $1 \leq t \leq \theta$.

2. Compute $e'_t = E_{pk_i}(b_{i,x,t}) - E_{pk_i}(b_{j,x,t}) = E_{pk_i}(\zeta'_t)$.

3. Compute $e''_t = (E_{pk_i}(b_{i,x,t}) - E_{pk_i}(b_{j,x,t}))^2 = E_{pk_i}(\zeta''_t)$.

4. Set $\gamma_0 = 0$, and compute $E_{pk_i}(\gamma_t)$ as $2E_{pk_i}(\gamma_{t-1}) + e''_t$, which implies $\gamma_t = 2\gamma_{t-1} + \zeta''_t$.

5. Select a random $r_t \in R_p$ in the form of $ax + b$ where $a, b \in \mathbb{Z}_p, a \neq 0$, and compute $E_{pk_i}(\delta_t)$ as $E_{pk_i}(\zeta'_t) + E_{pk_i}(r_t) \times (E_{pk_i}(\gamma_t) - E_{pk_i}(1))$, which implies $\delta_t = \zeta'_t + r_t(\gamma_t - 1)$.

6. Select a random $r_p \in \mathbb{Z}_p$ $(r_p \neq 0)$, and compute $E_{pk_i}(\mu_t)$ as

$$\sum_{h=1}^{\lambda}((s_{1,h} + s_{0,h})E_{pk_i}(1) + s_{1,h}E_{pk_i}(\delta_t) - s_{0,h}E_{pk_i}(\delta_t))$$

$$\times (r_p((E_{pk_i}(v_h))^2 - E_{pk_i}(v_h)) + E_{pk_i}(v_h)) \qquad (2.3)$$

$$+ r_p(\sum_{h=1}^{\lambda} E_{pk_i}(v_h) - E_{pk_i}(1)).$$

which implies $\mu_t = \sum_{h=1}^{\lambda}(s_{1,h}(1 + \delta_t) + s_{0,h}(1 - \delta_t))((v_h^2 - v_h)r_p + v_h) + (\sum_{h=1}^{\lambda} v_h - 1)r_p$.

Then, u_j compiles $E_{pk_i}(\mu) = (E_{pk_i}(\mu_1), \cdots, E_{pk_i}(\mu_\theta))$, and makes a random permutation to obtain $d_j = \mathscr{P}(E_{pk_i}(\mu))$. It finally sends a 5-tuple $(pid_j, cert_{pid_j}, a_x, d_j, Sign_{psk_j}(a_x, d_j))$ to u_i.

Step 3. u_i checks the validity of the received 5-tuple. Then, it decrypts every ciphertext $E_{pk_i}(\mu_t)$ in d_j as follows: for $E_{pk_i}(\mu_t) = (c_0, \cdots, c_\alpha)$, obtain μ_t by $\mu_t = (\sum_{h=0}^{\alpha} c_h s^h) \bmod p$. If $v_{i,x} > v_{j,x}$, u_i is able to find a plaintext $\mu_t \in \mathbb{Z}_p$ and $\mu_t = 2s_{1,y} \leq p - 1$ and computes $s_{1,y}$; if $v_{i,x} < v_{j,x}$, u_i is able to find $\mu_t = 2s_{0,y}$ and computes $s_{0,y}$.

Effectiveness Discussion: The correctness of the iCPM can be verified as follows. If $v_{i,x} > v_{j,x}$, then there must exist a position, say the t^*-th position, in the binary expressions of $v_{i,x}$ and $v_{j,x}$ such that $b_{i,x,t^*} = 1, b_{j,x,t^*} = 0$ and $b_{i,x,t'} = b_{j,x,t'}$ for all $t' < t^*$. Since $\gamma_t = 2\gamma_{t-1} + \zeta''_t$, we have $\gamma_{t'} = 0$, $\gamma_{t^*} = 1$, and $\delta_{t^*} = 1$. For $t'' > t^*$, we have $\gamma_{t''} \geq 2$, and δ_t is a random value due to $r_{t''}$. Since $s_{0,y}$ and $s_{1,y}$ are elements of $\mathbb{Z}_p$ and r_t is in the form of $ax + b$ $(a, b \in \mathbb{Z}_p, a \neq 0)$, u_i can always determine the effective plaintext from others. The effective plaintext will be $\mu_t = \sum_{h=1}^{\lambda}(s_{1,h}(1 + \delta_{t^*}) + s_{0,h}(1 - \delta_{t^*}))((v_h^2 - v_h)r_p + v_h) + (\sum_{h=1}^{\lambda} v_h - 1)r_p$. If the vector V_i from u_i does not satisfy $\sum_{h=1}^{\lambda} v_h = 1$ or $v_h \in \{0, 1\}$, u_i cannot remove the random factor r_p; if V_i satisfies the conditions, only $s_{1,y}$ and $s_{0,y}$ will be involved in the computation. Because $\delta_{t^*} = 1$, u_i can obtain $\mu_t = 2s_{1,y} \leq p - 1$ and recovers $s_{1,y}$. If $v_{i,x} < v_{j,x}$, we similarly have $\mu_t = 2s_{0,y}$ and u_i can obtain $s_{0,y}$.

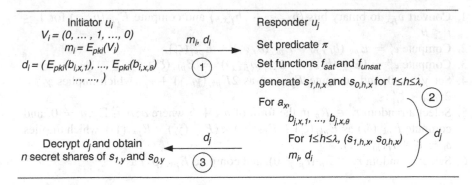

Fig. 2.3 The iPPM flow

The confidentiality of user profiles is guaranteed by the homomorphic encryption. The comparison result δ_t is always in the encrypted format, and δ_t is not directly disclosed to u_i. The revealed information is either $s_{1,y}$ or $s_{0,y}$ which is unrelated to user profiles. Therefore, the protocol transactions do not help in guessing the profiles, and the full anonymity is provided. In the meantime, vector V_i is always in an encrypted format so that u_j is unable to know the interested category T_y of u_i. In addition, u_j ensures that only one of $s_{1,y}$ and $s_{0,y}$ will be revealed to u_i. The non-forgeability property is similar to that of the eCPM. u_i will not lie as it makes signature $Sign_{psk_i}(a_x, d_i)$ and gives it to u_j. The profile forgery attack will be detected if u_j reports the signature to the TA. Moreover, u_j has no need to lie as it can achieve the same objective by simply modifying the contents of $s_{1,y}$ and $s_{0,y}$.

2.3.3 Approach 3: Implicit Predicate-Based Approach

Both the eCPM and the iCPM perform profile matching on a single attribute. For a matching involving multiple attributes, they have to be executed multiple times, each time on one attribute. In this section, we extend the iCPM to the multi-attribute cases, without jeopardizing its anonymity property, and obtain an implicit Predicate-based Profile Matching protocol, i.e., iPPM. This protocol relies on a predicate which is a logical expression made of multiple comparisons spanning distinct attributes and thus supports sophisticated matching criteria within a single protocol run (similar to [86]).

As shown in Fig. 2.3, the iPPM is composed of three main steps. In the first step, different from the iCPM, u_i sends to u_j n encrypted vectors of its attribute values corresponding to the attributes in A where A ($|A| = n \le w$) is the attribute set of the predicate Π. In the second step, u_j sets 2λ polynomial functions $f_{sat,h}(x)$, $f_{unsat,h}(x)$ for $1 \le h \le \lambda$. u_j then generates $2\lambda n$ secret shares from $f_{sat,h}(x)$, $f_{unsat,h}(x)$ by choosing $1 \le h \le \lambda, 1 \le x \le n$, and arranges them in a certain structure according to the predicate Π. For every 2λ secret shares with the same index h, similar to the Step 2 of the iCPM, u_j generates θ ciphertexts. u_j obtains $n\theta$ ciphertexts at the end of the second step. In the third step, u_i decrypts these $n\theta$ ciphertexts and finds n secret shares of $s_{1,y}$ and $s_{0,y}$. u_j finally can obtain $s_{1,y}$ or $s_{0,y}$ from the secret shares.

The iPPM is obtained by combining the iCPM with a secret sharing scheme [81] to support a predicate matching. The initiator u_i sends its attribute values corresponding to the attributes in A to the responder u_j. Without loss of generality, we assume $A = \{a_1, \cdots, a_n\}$. Then, u_j defines a predicate $\Pi = $ "$\bar{t}$ of $\{(v_{i,x}, opt, v_{j,x}) | a_x \in A\}$", where the comparison operator opt is either $>$ or $<$ and $\bar{t} \leq n$. The predicate contains n number of requirements (i.e., comparisons), each for a distinct a_x. The responder u_j determines λ pairs of messages $(s_{0,h}, s_{1,h})$ for attributes a_h ($1 \leq h \leq \lambda$). The initiator u_i receives $s_{1,h}$ if at least $\bar{t}$ of the n requirements are satisfied, or $s_{0,h}$ otherwise. Similar to the iCPM, T_y is determined by u_i but unknown to u_j. The threshold gate $1 \leq \bar{t} \leq n$ is chosen by u_j. When $n = 1$, the iPPM reduces to the iCPM. The protocol steps are given below.

Step 1. u_i generates a vector $V_i = (v_1, \cdots, v_\lambda)$, where $v_y = 1$ and $v_h = 0$ for $1 \leq h \leq \lambda$ and $z \neq y$, and sets $m_i = E_{pk_i}(V_i) = (E_{pk_i}(v_1), \cdots, E_{pk_i}(v_\lambda))$. In addition, u_i selects the attribute set A ($|A| = n$), and sends u_j a 6-tuple:

$$(pid_i, cert_{pid_i}, A, d_i, m_i, Sign_{psk_i}(A, d_i, m_i)),$$

where d_i contains $n\theta$ ($\theta = \lceil \log l \rceil$) ciphertexts as the homomorphic encryption results of each bit of $v_{i,x}$ for $a_x \in A$.

Step 2. u_j checks the validity of the received 6-tuple (similar to the Step 2 of the iCPM). It creates a predicate Π and chooses the threshold gate $\bar{t}$. Using the secret sharing scheme [81], u_j creates 2λ polynomials: $f_{sat,h}(v) = \rho_{\bar{t}-1,h} v^{\bar{t}-1} + \cdots + \rho_{1,h} v + s_{1,h}$ and $f_{unsat,h}(v) = \rho'_{n-\bar{t},h} v^{n-\bar{t}} + \cdots + \rho'_{1,h} v + s_{0,h}$ for $1 \leq h \leq \lambda$, where $\rho_{\bar{t}-1,h}, \cdots, \rho_{1,h}, \rho'_{n-\bar{t},h}, \cdots, \rho'_{1,h}$ are random numbers from $\mathbb{Z}_p^*$. For each attribute $a_x \in A$, it calculates the secret shares of $s_{1,h,x}$ and $s_{0,h,x}$ as follows ($s_{1,h,x}, s_{0,h,x} \leq (p-1)/2$ are required):

$$\begin{cases} s_{0,h,x} = 0 || f_{unsat,h}(x), \\ s_{1,h,x} = 1 || f_{sat,h}(x), & \text{if "} v_{i,x} > v_{j,x} \text{"} \in \Pi; \\ s_{0,h,x} = 1 || f_{sat,h}(x), \\ s_{1,h,x} = 0 || f_{unsat,h}(x), & \text{if "} v_{i,x} < v_{j,x} \text{"} \in \Pi. \end{cases} \quad (2.4)$$

Note that u_j adds a prefix 0 or 1 to each secret share such that u_i is able to differentiate the two sets of shared secrets, one for $s_{1,h}$, the other for $s_{0,h}$. u_j runs the Step 2 of the iCPM n times, each time for a distinct attribute $a_x \in A$ and with $(s_{1,h,x}, s_{0,h,x})$ for ($1 \leq h \leq \lambda$) being input as $s_{1,h}$ and $s_{0,h}$, respectively. u_j then obtains d_j including $n\theta$ ciphertexts. Finally, it sends a 6-tuple $(pid_j, cert_{pid_j}, \bar{t}, A, d_j, Sign_{psk_j}(d_j))$ to u_i.

Step 3. u_i checks the validity of the received 6-tuple. u_i can obtain n secret shares, and each of these shares is either for $s_{0,y}$ or $s_{1,y}$. It then classifies the n shares into two groups by looking at the starting bit (either "0" or "1"). Thus, if Π is satisfied, u_i can obtain at least $\bar{t}$ secret shares of $s_{1,y}$ and be able to recover $s_{1,y}$; otherwise, it must obtain at least $n - \bar{t} + 1$ secret shares of $s_{0,y}$ and can recover $s_{0,y}$.

The correctness of the iPPM is as follows. At Step 2, the responder u_j executes the Step 2 of the iCPM n times, each time it effectively delivers only one secret share of either $s_{0,y}$ or $s_{1,y}$ to u_i. When u_i receives either $\bar{t}$ shares of $s_{1,y}$ or $n - \bar{t} + 1$ shares of $s_{0,y}$, it can recover either $s_{1,y}$ or $s_{0,y}$. The interpolation function corresponding to the secret sharing scheme always guarantees the correctness. The anonymity and non-forgeability of the iPPM are achieved similar to those of the iCPM and the eCPM, respectively.

Efficiency Discussion: Let $|R|$ be the size of one ring element in R_q. In the eCPM, the initiator and the responder both need to send ciphertexts in size of $2|R|$, and the communication overhead is thus subject only to the system parameter $|R|$.

In order to achieve full anonymity, the iCPM constructs ciphertext in a sequence of operations. From Sect. 1.4.4.5, we know $|Enc(b)| = 2|R|$. Thus, the communication overhead of the initiator is $2(\theta + \lambda)|R|$ with $\theta = \lceil \log l \rceil$. It can be seen that the initiator's communication overhead increases with system parameters (θ, λ). According to Sect. 1.4.4.5 an addition operation of homomorphic encryption does not increase the ciphertext size, while a multiplication with inputs of two ciphertexts of lengths $a|R|$ and $b|R|$ outputs a $(a+b-1)|R|$-length ciphertext. Thus, in the iCPM, the communication overhead of the responder increases to $6\theta|R|$. It is concluded that the communication overhead of the eCPM and the iCPM is constantly dependent on system parameters (θ, λ).

The iPPM extends the iCPM by building complex predicates. From the protocol description, we observe that if a predicate includes $n \geq 1$ comparisons, the communication overhead of the iPPM would be approximately n times of that in the iCPM.

2.4 Anonymity Enhancing Techniques

2.4.1 Anonymity Measurement

Suppose that user u_i is currently using pseudonym $pi\,d_i$ to execute profile matching with others. We consider an adversary aiming to break the *k-anonymity* of u_i. We have the following definition:

Definition 2.4. The k-anonymity risk level of a user is defined as the inverse of the *minimum number of distinct protocol runs (MNDPR)* that are required to break the user's k-anonymity.

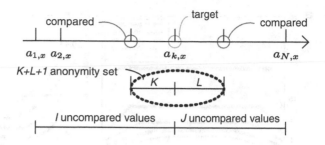

Fig. 2.4 Identifying the target from others

From this definition, the k-anonymity risk level reflects the difficulty that the adversary can break a user's k-anonymity. In the iCPM and the iPPM, the profile matching initiator does not reveal its attribute values to the responder, and the responder has no clue about the comparison result and only reveals the self-defined messages which are not related to the profile. In this case, a user's k-anonymity risk level can be minimum, i.e., no matter how many protocol runs are executed, its k-anonymity risk level is always the lowest. Therefore, the iCPM and the iPPM both provide full anonymity (put users at minimum anonymity risk).

For the eCPM, it exposes the comparison results to users and thus obviously puts users at risk of the disclosure of attribute values. Because every eCPM run is executed for a particular attribute (which is specified by the initiator), any user u_i has a k-anonymity risk level on its each individual attribute. When "=" case happens, users have higher anonymity level because they will be indistinguishable from other users with the same attribute values. In the following, we consider the worst case where users have distinctive attribute values on a single attribute. For a given attribute a_x, we assume $a_{1,x} > a_{2,x} > \cdots > a_{N,x}$, where $v_{i,x}$ is the value of u_i on a_x. In order to break u_i's k-anonymity on a_x, the adversary has to make comparisons "$a_{\alpha,x} > v_{i,x}$" and "$v_{i,x} > a_{\beta,x}$" for $\beta - \alpha - 1 < k$ so that the anonymity set of $v_{i,x}$ has a size smaller than k. Let I and J respectively be the numbers of larger and smaller values on a_x among all the users that have not been compared to $v_{i,x}$. Let $K \le I$ and $L \le J$ respectively be the number of such un-compared values in the k-anonymity set of $v_{i,x}$. The relations among I, J, K, and L are shown in Fig. 2.4. Assuming the contact is uniformly random, we define a recursive function f as shown in Eq. (2.5).

$$f(I,J,K,L) = \begin{cases} 0, & \text{if } K + L < k - 1 \text{ or } I < K \text{ or } J < L; \\[2mm] \dfrac{(I-K)(f(I-1,J,K,L)+1) + (J-L)(f(I,J-1,K,L)+1)}{I+J} + \\[2mm] \dfrac{\sum_{z=1}^{K}(f(I-1,J,K-z,L)+1) + \sum_{z=1}^{L}(f(I,J-1,K,L-z)+1)}{I+J}, & \text{otherwise} \end{cases} \tag{2.5}$$

The above function $f(I,J,K,L)$ returns the MNDPR with respect to a user's k-anonymity on a_x in the eCPM. Thus, the user's anonymity risk level in this case is defined as $\mathscr{L} = 1/f(I,J,K,L)$. Since we assumed $a_{1,x}, \cdots a_{N,x}$ are sorted in a descending order, the index i actually reflects the rank of $v_{i,x}$ among the attribute

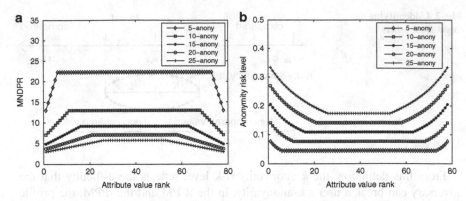

Fig. 2.5 Numerical results on user anonymity risk level. (**a**) MNDPR per 78 users. (**b**) Anonymity risk level

values. Figure 2.5 plots the MNDPR $f(I, J, K, L)$ and the k-anonymity risk level $\mathscr{L}$ in terms of 78 users' attribute values where $k = 5, 10, \cdots, 25$. It can be seen that a user with a median attribute value will have a lower k-anonymity risk level than those with larger or smaller values. This is reasonable because the user with a median attribute value is less distinctive from other users.

2.4.2 Anonymity Enhancement

We have derived the maximum number of distinct eCPM runs (MNDPR) before a user's k-anonymity is broken. This number is obtained under an assumption of uniformly random contact. However, in reality, users as social entities are likely to gather with others who have similar attribute values. This situation increases user anonymity risk level quickly when profile matching is executed frequently, and the k-anonymity can be broken within a much smaller number of the eCPM runs as a result. Recall that multi-pseudonym techniques are used to protect user identity and location privacy. Similar to previous work [21, 75], here we consider that pseudonyms themselves are unlinkable. In the eCPM, if a user does not change the pseudonym, the comparison result will be easily linked to break the k-anonymity. If a user changes pseudonym for each protocol run, information revealed by the protocol cannot be directly linked, and the user obtains highest anonymity. Nevertheless, it is desirable that the user changes pseudonym only when necessary, since pseudonyms are limited resources and have associated cost [21, 75] (e.g., communication cost for obtaining them from the TA and computation cost for generating them on the TA). Thus, user anonymity is tightly related with pseudonym change frequency.

Our goal is to improve the anonymity strength of the eCPM by combining it with a pre-adaptive pseudonym change strategy which enables users to take necessary pseudonym change action before their k-anonymity is broken. The new verso of the eCPM is referred to as eCPM+. Before presenting the pre-adaptive strategy, we first introduce a post-adaptive pseudonym change strategy, where users measure their anonymity risk levels periodically and change their pseudonym after their anonymity risk levels becomes larger than a pre-defined threshold value.

The post-adaptive strategy assumes that a user u_j as responder runs the protocol on an attribute a_x with an initiator u_i (recognized by seeing the same pseudonym) only once, and refuses to participate any subsequent protocol running on the same a_x with u_i. However, if u_i has changed its pseudonym since the last protocol running with u_j, then u_j will consider u_i as a *new partner* and participate the protocol. Time is divided into slots of equal duration. The *neighborhood status* of u_i on attribute a_x in a time slot is characterized by a pair of values $NS_{i,x} = (n_{i,x,s}, n_{i,x,l})$, respectively implying the number of new partners (identified in the time slot) with attribute values smaller than $v_{i,x}$ and the number of those with attribute values larger than $v_{i,x}$. It varies over time due to user mobility and can be modeled as a time series data.

The centre of this strategy is the continuous measurement of user anonymity risk level based on neighborhood status. In the iCPM, attribute values are protected, and users obtain the matching results. For every attribute a_x, user u_i maintains the numbers $N_{i,x,s}$ and $N_{i,x,l}$ of discovered values that are smaller and larger than its own value $v_{i,x}$ since the last change of pseudonyms. These two numbers are respectively the sum of individual $n_{i,x,s}$ and the sum of $n_{i,x,l}$ corresponding to the past several time slots. Recall that $v_{i,x}$ is ranked the i-th largest among all N users in the network. Let $I = i - 1$ and $J = N - i$. u_i is not able to compute the accurate MNDPR because it does not have the information of the last two arguments of function $f()$ (see Eq. (2.5)). The anonymity risk level of u_i on a_x may be estimated as $\mathscr{L} = 1/f'(N_{i,x,s}, N_{i,x,l})$, where $f'(N_{i,x,s}, N_{i,x,l})$ approximates the MNDPR of u_i regarding a_x and is given as $\sum_{\substack{1 \le \alpha \le I - N_{i,x,s} \\ 1 \le \beta \le J - N_{i,x,l}}} \Pr[(\alpha, \beta)] \cdot f(I - N_{i,x,s}, J - N_{i,x,l}, \alpha, \beta)$.

For simplicity, we assume that the $N_{i,x,s}$ values are randomly distributed among the $I - \alpha$ users ($0 \le \alpha \le I - N_{i,x,s}$) with larger values on a_x than u_i and the $N_{i,x,l}$ values are randomly distributed among the $J - \beta$ smaller-value users ($0 \le \beta \le J - N_{i,x,l}$). Thus, for $N_{i,x,s} \ge 1$ and $N_{i,x,l} \ge 1$, we have $f'(N_{i,x,s}, N_{i,x,l})$ as

$$\sum_{\substack{0 \le \alpha \le I - N_{i,x,s} \\ 0 \le \beta \le J - N_{i,x,l}}} \frac{\binom{I - \alpha - 1}{N_{i,x,s} - 1}\binom{J - \beta - 1}{N_{i,x,l} - 1}}{\binom{I}{N_{i,x,s}}\binom{J}{N_{i,x,l}}} f(I - N_{i,x,s}, J - N_{i,x,l}, \alpha, \beta). \qquad (2.6)$$

For $N_{i,x,s} = 0$ and $N_{i,x,l} \ge 1$, $f'(N_{i,x,s}, N_{i,x,l})$ is

$$\sum_{0 \le \beta \le J - N_{i,x,l}} \frac{\binom{J - \beta - 1}{N_{i,x,l} - 1}}{\binom{J}{N_{i,x,l}}} \cdot f(I, J - N_{i,x,l}, I, \beta). \qquad (2.7)$$

For $N_{i,x,s} \geq 1$ and $N_{i,x,l} = 0$, $f'(N_{i,x,s}, N_{i,x,l})$ is

$$\sum_{0 \leq \alpha \leq I - N_{i,x,s}} \frac{\binom{I-\alpha-1}{N_{i,x,s}-1}}{\binom{I}{N_{i,x,s}}} \cdot f(I - N_{i,x,s}, J, \alpha, J). \qquad (2.8)$$

In the above computation, u_i needs to know N and its value rank i. The information can be obtained from the TA when u_i registers to the TA. If users are allowed to freely leave and enter the network, they will need to de-register/re-register themselves with the TA when leaving/joining the network. In this case, (N, t) are changing, and the TA has to be involved in the network operation in order to maintain latest network status and update users with the latest information.

The post-adaptive strategy also relies on *pseudonym lifetime* for making pseudonym change decisions. Suppose that user u_i is currently using pseudonym $pi d_i$. The longer $pi d_i$ has been used, the more private information of u_i is leaked in case its anonymity has been broken. Hence, when u_i's anonymity risk level $\mathcal{L}_i$ has stayed unchanged for a certain duration, called the lifetime of $pi d_i$ and denoted by $\tau(pi d_i)$, u_i changes its pseudonym for damage control. However, $\tau(pi d_i)$ should not be given as a constant value, but subject to $\mathcal{L}_i$. The higher $\mathcal{L}_i$ is, the more possible the anonymity of u_i is broken, and therefore the smaller $\tau(pi d_i)$ is. We define $\tau(pi d_i) = \xi \frac{\text{MNDPR}_i}{\mathcal{L}_i}$, where MNDPR_i is obtained by Eq. (2.5) and $\xi > 1$ is the pseudonym lifetime factor.

For the pre-adaptive pseudonym change strategy, each user u_i initializes an ARMA model for its neighborhood status on every attribute when entering the network. Since it has w attributes, the number of ARMA models to be initialized is w. At the end of each time slot, it measures its current neighborhood status on each attribute and updates the corresponding ARMA models. It takes the post-adaptive strategy for each attribute to determine whether to change its pseudonym. In case pseudonym change is not suggested, it proceeds to predict the neighborhood status on all the attributes in the following time slot using the ARMA models. If one of the predicted neighborhood status leads to an unacceptable anonymity risk level, it changes its pseudonym; otherwise, it does not. The pre-adaptive strategy strengths the post-adaptive strategy by one-step ahead prediction based decision making and generally enhances user anonymity.

2.5 Performance Evaluation

The eCPM+ addresses accumulative anonymity risk in multiple protocol runs and tunes itself automatically to maintain desired anonymity strength. Some previous works [24, 28] are concerned only with the anonymity risk brought by each individual protocol run, and some works [22] reduce anonymity risk by manually adjusting certain threshold values. Though they provide the conditional anonymity as the eCPM, they are not comparable to the eCPM and the eCPM+ because the anonymity protection of users is considered in terms of consecutive protocol runs. Therefore, in this section we evaluate the eCPM+ (which uses a pre-adaptive pseudonym change strategy) in comparison with two other eCPM variants, respectively employing a constant pseudonym change interval z (CONST-z) and a post-adaptive pseudonym change strategy (Post).

2.5.1 Simulation Setup

Our simulation study is based on the real trace [32] collected from 78 users attending a conference during a 4-day period. A contact means that two users come close to each other and their attached Bluetooth devices detect each other. The users' Bluetooth devices run a discovery program every 120 s on average and logged about 128,979 contacts. Each contact is characterized by two users, a start-time, and a duration. In CONST-z, we set the pseudonym change interval z from 1 to 40 (time slots); in the post-adaptive and pre-adaptive strategies, we set pseudonym lifetime factor $\xi = 30$. In the pre-adaptive strategy, we use ARMA order $(10, 5)$.

We use the contact data to generate user profiles. According to social community observations [87], users within the same social community often have common interests and are likely interconnected through strong social ties [16]. The stronger tie two users have, the more likely they contact frequently. Let $f_{i,j}$ denote the number of contacts of users u_i and u_j. We build a complete graph of users and weight each edge (u_i, u_j) by $f_{i,j}$. By removing the edges with a weight smaller then 100, we obtain a graph G containing 78 vertices and 2,863 edges. We find all maximal cliques in G using the Bron–Kerbosch algorithm [88]. A clique is a complete subgraph. A maximal clique is a clique that cannot be extended by including one more adjacent vertex. We obtain the 7,550 maximal cliques $C_1, \cdots, C_{7550}$ that all contain ≥ 15 users.

Without loss of generality, we assume that these cliques are sorted in the descending order of the weight sum of their edges (the weight sum of C_1 is the largest). We then construct communities in the following way. Scan the sequence of cliques from C_1 to C_{7550}. For a scanned clique C_i, find a clique C_j that has been previously scanned and identified as *core clique* and contains $\geq 80\%$ vertices of C_i. If there are multiple such cliques, take the one with largest weight sum as C_j. If C_j is found, assign C_i with the same attribute as C_j; otherwise, generate a new attribute, assign it to C_i, and mark C_i as a core clique. After the attribute generation and assignment, merge the cliques with the same attribute into a community. A community contains multiple users, and a user may belong to multiple communities. From the above settings, we generate 349 attributes and thus obtain 349 communities. We however concentrate on the first generated 100 attributes and their corresponding communities for simplicity. On average, each of these considered communities contains 28 users, and each user belongs to 38 communities.

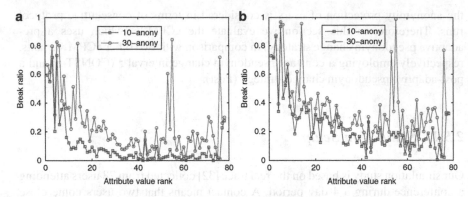

Fig. 2.6 Anonymity break period under the constant strategy. (**a**) $z = 1$. (**b**) $z = 20$

Afterwards, we assign values to each user in G for these 100 attributes. For an attribute a_x, we find the corresponding community $\mathscr{C}_x$ and do the following. For each user in $\mathscr{C}_x$, we compute the weight sum of its incidental edges in $\mathscr{C}_x$; for each vertex outside $\mathscr{C}_x$, we compute the weight sum of its incident edges to the vertices in $\mathscr{C}_x$; then, we sort all the users in the decreasing order of their weight sums and assigned their values on a_x with $(78, 77, \cdots, 1)$. This assignment method is reasonable because a large weight sum indicates a large interest in communicating with users in $\mathscr{C}_x$ and thus a strong background in the aspect represented by a_x.

Our simulation spans 10,000 time slots, each lasting 30 s, and focuses on a randomly selected attribute. Users can change their pseudonym at the beginning of each time slot. The pseudonym is *corrupted* in terms of k-anonymity (on the selected attribute) if there are less than $k - 1$ other users in the network that will obtain the same matching results in the same protocol settings. A user experiences an anonymity break (on the selected attribute) if it is using a *corrupted* pseudonym.

2.5.2 Simulation Results

Figure 2.6 shows the anonymity break period experienced by each user with the constant strategy being used. It can be seen that when $z = 1$, each user experiences the shortest anonymity break period at the cost of 10,000 pseudonyms per user. Anonymity break is still possible in this extreme case because users may have multiple contacts within a single time slot while they are still using the same pseudonym. If a user has a more restrictive anonymity requirement (e.g., from 10-anonymity to 30-anonymity) or uses a larger pseudonym change interval (from 1 time slot to 20 time-slots), it will have more *corrupted* pseudonyms and thus suffer a longer period of anonymity break.

The neighborhood status of a user on a given attribute is characterized by the number of neighbors with larger values and the number of neighbors with smaller

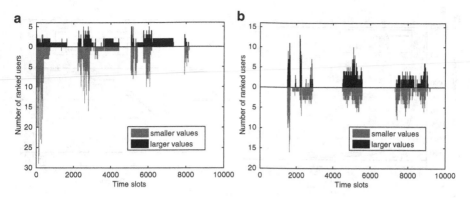

Fig. 2.7 Neighborhood status over time. (**a**) The 7th user. (**b**) The 32nd user

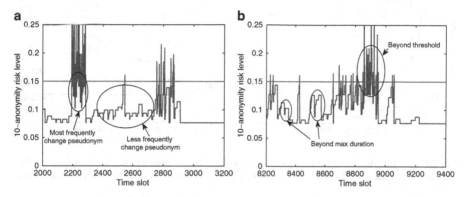

Fig. 2.8 Anonymity risk level over time ($th = 0.15$). (**a**) Time period $(2000, 3200)$. (**b**) Time period $(8200, 9400)$

values. We investigate the regularity of neighborhood status of individual users over time and justify the effectiveness of pre-adaptive strategy. To do so, we randomly choose two users, ranked respectively the 7th and the 32nd. Figure 2.7 shows their neighborhood status. The 7th user's neighborhood status exhibits regular change, i.e., the number of neighbors with larger values stays stable, and that of neighbors with smaller values decrease linearly over time. For the 32nd user, the number of users with larger values and the number of users with smaller values both decrease.

We choose the 32nd user, who in general has lower anonymity risk level than the 7th user, and show its 10-anonymity risk level in two consecutive time periods $(2000, 3200)$ and $(8200, 9400)$ with the post-adaptive strategy in Fig. 2.8. The anonymity risk level threshold is $th = 0.15$. In the figure, the drop from a high risk level to a low risk level indicates a pseudonym change. Recall that a user changes its pseudonym not only when the anonymity risk level is beyond threshold th but also when its current pseudonym expires. This is reflected by the anonymity risk level drop happened below the threshold line in the figure. From Fig. 2.7, we can

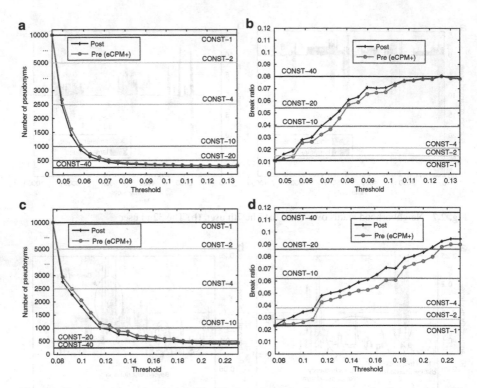

Fig. 2.9 Pseudonyms and break ratio (the 32nd user). (**a**) # of pseudonyms for 5-anonymity. (**b**) 5-anonymity break period. (**c**) # of pseudonyms for 10-anonymity. (**d**) 10-anonymity break period

see that the pseudonym change frequency is high when the user encounters a large number of neighbors. This is reasonable as a large number of profile matching runs are executed in this case, and the user's anonymity risk level grows quickly. When the level is beyond a pre-defined threshold, the user changes its pseudonym.

Figure 2.9 shows the performance of the constant, the post-adaptive and the pre-adaptive strategies respectively for 5-anonymity and 10-anonymity, in relation with threshold th. The results are obtained with respect to the 32nd user. For the constant strategy, multiple lines are plotted, respectively corresponding to $z = \{1, 2, 4, 10, 20, 40\}$. As z goes up, the user consumes a decreasingly number of pseudonyms and has an increasingly break ratio (the ratio of the number of time slots that the k-anonymity of the 32nd user is broken to 10,000). It can be seen that the number of pseudonyms consumed by the post-adaptive and pre-adaptive strategies are much smaller than those of the constant strategy. For example, in the case of 5-anonymity and $th = 0.0763$, the post-adaptive strategy spends 369 pseudonyms and results in a 514 time slot anonymity break period. The constant strategy consumes 500($>$369) pseudonyms and has a 0.0540($>$0.0514) break ratio. The post-adaptive strategy outperforms the constant strategy in anonymity protection by using fewer pseudonyms to achieve smaller break ratio. Similar phenomena are observed for

other th values and 10-anonymity scenario as well. In particular, we find that as
expected, the pre-adaptive strategy leads to yet better anonymity performance than
the post-adaptive one. Figure 2.9 shows that in case of 5-anonymity and $th =$
0.0763, the pre-adaptive strategy consumes 449($>$369) pseudonyms and results in
a 0.0445($<$0.0514) break ratio. The pre-adaptive strategy consumes slightly more
pseudonyms, but achieves significantly shorter anonymity break period.

2.6 Related Work

In general, the profile matching can be categorized based on the formats of profiles
and the types of matching operations. A well-known profile matching is the FNP
scheme [68], where a client and a server compute their intersection set such that
the client gets the result while the server learns nothing. Later, Kissner and Song
[89] implemented profile matching with more operations including set intersection,
union, cardinality and over-threshold operations. On the other hand, Ye et al. [90]
further extended the FNP scheme to a distributed private matching scheme and
Dachman-Soled et al. [91] aimed at reducing the protocol complexity. All the above
solutions to the set intersection rely on homomorphic encryption operation. In the
meantime, other works [92, 93] employed an oblivious pseudo random function to
build their profile matching protocols, where communication and computational
efficiency is improved. Li et al. [24] implemented profile matching according to
three increasing privacy levels: (1) revealing the common attribute set of the two
users; (2) revealing the size of the common attribute set; and (3) revealing the size
rank of the common attribute sets between a user and its neighbors. They considered
an honest-but-curious (HBC) adversary model, which assumes that users try to learn
more information than allowed by inferring from the profile matching results, but
honestly following the protocol. They applied secure multiparty computation, the
Shamir secret sharing scheme, and the homomorphic encryption scheme to achieve
the confidentiality of user profiles.

   In another category of profile matching [22, 28, 94], profiles can be represented
as vectors, and matching operation can be inner product or distance. Such profile
matching is a special instance of the secure two-party computation, which was
initially introduced by Yao [95] and later generalized to the secure multi-party
computation by Goldreich et al. [96]. Specifically, we introduce two recent works
in this category. Dong et al. [22] considered user profile consisting of attribute
values and measured the proximity of two user profiles using dot product $f_{dot}(u, v)$.
An existing dot product protocol [97] is improved to enable verifiable secure
computation. The improved protocol only reveals whether the dot product is above
or below a given threshold. The threshold value is selected by the user who
initiates the profile matching. They pointed out the potential anonymity risk of
their protocols; an adversary may adaptively adjust the threshold value to quickly
narrow down the value range of the victim profile. Thus, it is required that the
threshold value must be larger than a pre-defined lower bound (a system parameter)

to guarantee user anonymity. The same problem exists in other works [24, 28]. Furthermore, Dong et al. [22] required users to make a commitment about their profiles to ensure the profile consistency, but profile forgery attack may still take place during the commitment phase. In the same category, Zhang et al. [28] set the matching operation $f_{dis}(u, v)$ of two d-dimension user profiles u and v as the calculation of the following distances: (1) Manhattan distance, i.e., $f_{dis}(u, v) = l_\alpha(u, v) = (\sum_1^d |v_i - u_i|^\alpha)^{\frac{1}{\alpha}}$; or (2) Max distance, i.e., $f_{dis}(u, v) = l_{max}(u, v) = \max\{|v_1 - u_1|, \cdots, |v_d - u_d|\}$. The distance is compared with a pre-defined threshold τ to determine whether u and v match. Then, three increasing privacy levels are defined as: (1) one of u and v learns $f_{dis}(u, v)$, and the other only learns f_{dis}; (2) one of them learns $f_{dis}(u, v)$, and the other learns nothing; and (3) one of them learns whether $f_{dis}(u, v) < \tau$, and the other learns nothing.

2.7 Conclusion and Future Directions

We have investigated the privacy-preserving profile matching problem, and introduced three comparison-based profile matching protocols including the explicit comparison-based profile matching protocol (eCPM), the implicit comparison-based profile matching protocol (iCPM), and the implicit predicate-based profile matching protocol (iPPM). We have shown that the eCPM achieves *conditional anonymity* and both the iCPM and the iPPM achieves *full anonymity*. We have further introduced an enhanced version of the eCPM, i.e., eCPM+, by exploiting the prediction method and the pre-adaptive pseudonym change. The effectiveness of the eCPM+ is validated through extensive simulations using real-trace data.

In the research of profile matching, matching operations, matching results, and user anonymity are three adjustable parameters. The design goal of profile matching protocols is always to find a tradeoff between these metrics and useability. The more anonymity-level achieved, the less information users want to share. In this book, we have summarized the related profile matching protocols, indicated their conditional anonymity property, and introduced the protocols with full anonymity, which will be very useful for mobile applications with high privacy requirements. The existing open problems in the privacy-preserving profile matching research are shown below:

- Design profile matching with more primitive operations, such as "$\geq$" and "$\leq$". The eCPM, iCPM and iPPM implement "$>$" and "$<$" as the matching operation. How to implement "$\geq$" and "$\leq$" matching operations is still an open problem. The anonymity analysis of "$=$" case is also interesting. The larger number of users that have the same dimensional value, the higher anonymity level users would obtain. This is because they are indistinguishable from others who have the same dimensional value. In the eCPM, if two users have the same dimension values $v_{i,x} = v_{j,x}$, they would be able to identify that fact and their anonymity will be reduced to the minimum level. To study the impact of "$=$" on the anonymity is a significant future research direction.

- Study the anonymity variance in different environment conditions. The anonymity is significantly different depending on the application scenarios. For example, if a user walks in a conference room where people around have similar interests (implying similar dimensional values), its anonymity would decrease quickly. This is because the range of its profiles can be quickly narrowed down. On the other hand, if a user visits a plaza where people around have no common interests, its anonymity would probably decrease slowly. We demonstrate the anonymity variance in a simulation built based on the real-trace from a conference. It is of significance to study the profile matching protocols in different environments and evaluate the performance by choosing different pseudonym change strategies.

Chapter 3
Cooperative Data Forwarding Strategy with Privacy Preservation

3.1 Introduction

In the MSN, users rely on the opportunistic contacts for cooperative data forwarding. Unlike conventional wireless relay networks assuming end-device to be insensate, users have specific social features, e.g., privacy concerns and selfishness, and they will choose a data forwarding strategy under the impacts of these features. For example, when privacy preservation of users is applied, users become unrecognizable to each other and the social ties and interactions are no longer traceable. In this case, there is no obvious incentives for users to be cooperative on data forwarding. Thus, due to the selfish behavior, the cooperative data forwarding could be severely interrupted or even disabled.

In this chapter, we address the privacy preservation problem and the data forwarding problem in the MSN. Our goal is to resolve the two problems in one framework by proposing a privacy preserving social-based cooperative data forwarding protocol. We exploit the social morality for cooperative data forwarding design. The morality of human beings is a common social phenomenon which provides the rules for people to act upon and grounds the moral imperatives. It is the fundament of a cooperative and mutually beneficial social life in the real-world society. Specifically, we will address the problem according to three steps.

First, we identify the conflicting nature between privacy preservation and cooperative data forwarding in the MSN. We leverage social morality to model the user cooperation and accordingly promote the communication efficiency.

Second, we introduce a three-step protocol suite to attain the privacy-preserving data forwarding. In step one, we introduce a privacy-preserving route-based authentication scheme. It enables users to expose the mobility information to each other for cooperation, yet with location privacy preserving. In step two, based on the mobility of users, we evaluate the forwarding capability of individual users on a given packet. In step three, a game-theoretic approach taking account of both the morality and forwarding capability is designed to adaptively determine the optimal data forwarding strategy for individual users.

X. Liang et al., *Security and Privacy in Mobile Social Networks*, SpringerBriefs in Computer Science, DOI 10.1007/978-1-4614-8857-6_3, © The Author(s) 2013

Third, we evaluate the performance of the protocols through extensive trace-based simulations. The simulation results validate the efficiency of the data forwarding protocols and the location privacy preservation.

The remainder of this chapter is organized as follows: in Sect. 3.2, we present the network model and design goal. Then, we introduce the three-step protocol suite in Sect. 3.3. We evaluate the protocol in a trace-based simulation environment in Sect. 3.4. Lastly, we review the related work and draw our summary respectively in Sects. 3.5 and 3.6.

3.2 Models and Design Goal

3.2.1 Network Model and Social Behavior Model

We consider a homogenous MSN composed of a set $\mathcal{V} = \{u_1, \cdots, u_N\}$ of mobile users with the network size $|\mathcal{V}| = N$. Users have equal communication range, denoted by R_t. The communication between any two users u_i and u_j, is bidirectional, i.e., u_i can communicate to user u_j if and only if user u_j can also communicate to u_i. Users follow the same behavior model: they are selfish, tending to maximize their individual utilities during data forwarding, and do not perform irrational attacks. A trusted authority (TA) is available at the initialization phase for generating pseudonyms and secret keys for MSN users, but it will not be involved in the data forwarding. Users continuously change their pseudonyms to preserve their identity and location privacy. The pseudonym change breaks any relation previously established between two users and as a result they can no longer recognize each other.

User-to-Spot Data Forwarding: We assume that there exists a set $\mathcal{A} = \{a_1, \cdots, a_l\}$ of social hotspots in the network. They are located in regions such as supermarkets, restaurants, office buildings and residential blocks with high population density as shown in Fig. 1.1. Different users have different sets of favored hotspots that they frequently visit. The hotspots that a user visited in the past indicate the personal preference of the user and thus may relate to the user's future locations [98]. In addition, the hotspots can be categorized into sensitive hotspots, e.g., office buildings, residential blocks, and non-sensitive hotspots, e.g., supermarkets, restaurants. Sensitive hotspots are tightly related to users' private lives. The access to sensitive hotspots needs to be protected according to users' privacy needs. In this work, we apply the hotspot technique [99] to preserve receiver location privacy.

We introduce a user-to-spot data forwarding protocol to achieve privacy preservation and user cooperative data forwarding. Specifically, each hotspot is equipped with a non-compromised and communicable storage device which buffers the packets for receivers to fetch. A data sender/forwarder leaves packets at selected hotspots, and receivers can fetch the packets upon their later access to the same hotspots. Compared with the contact-based data forwarding protocols where users

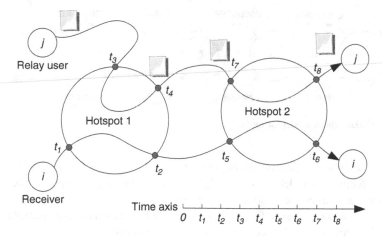

Fig. 3.1 An illustration of effective data forwarding

swap data upon their contacts, the user-to-spot data forwarding protocol would have more successful deliveries in special cases as shown in Fig. 3.1. In this figure, relay user u_j has no contact with receiver u_i but they enter the common hotspots during different time periods. By making use of this property, the user-to-spot data forwarding protocol enables u_j to deliver the packet to u_i. This user-to-spot data forwarding protocol is practical due to the following facts:

- Social users often have specific preferences on common social hotspots, such as supermarkets, office buildings, etc. They are likely to choose part of these hotspots and visit them frequently.
- In the MSN, data sender often has certain social relationship with receiver. The sender is likely to be partially aware of the social behaviors and frequently-visited hotspots of the receiver.
- Hotspot buffers are low-cost and can be pervasively available data storage resources [100]. They are not interconnected and will not be involved in cooperative data forwarding. They act as static receivers to temporarily store user data and allow authorized wireless access of the data when users come into their wireless communication range.

In this work, the identity of the receiver is implicitly contained (thus protected) in the packet, and the receiver can fetch the packet from the hotspot buffer after a simple authentication operation, e.g., using the scheme in [99].

Observing the unique social features in the MSN, we exploit the morality factor of the MSN by mimicking the morality-centric human society. We emphasize that the morality factor should be counted into the calculation of users' utility. To this end, we instantiate two forms of social morality, i.e., guilt and high-mindedness, in the context of MSN-based data forwarding where cooperation is highly desirable: users feel *guilty* when they defect (i.e., refuse to forward a packet), and they feel

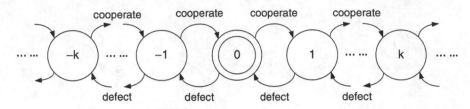

Fig. 3.2 Markov chain model for morality state

high-minded when choosing to cooperate (i.e., help to forward a packet). Guilt creates a feeling of indebtedness, which directs them to cooperate, while high-mindedness alleviates the guilty feeling of users.

A self-regulated morality factor g, internalized for each user that quantitatively depicts the internal moral force, is based on two elements:

- *Morality state x*: The morality state reflects the behavior history of a user. It increases by one level for a single cooperation behavior and decreases by one level due to a single defection conduct.
- *Sociality strength st*: The sociality strength st is related to a user's personal experience, such as education and habitation. It is stabilized and less independent with short-term behavior changes. If the sociality strength of a user is significant, the user feels a correspondingly significant increment of guilt towards a single defection behavior and a correspondingly significant increment of high-mindedness towards a single cooperation behavior.

Each u_i has a sociality strength denoted by st_i, and a varying morality state x_i. Following social theory [37, 38], we depict the morality state x_i by a Markov chain model with the state space and non-null transitions shown in Fig. 3.2. Let $P_i(j, j + 1)$ and $P_i(j, j - 1)$ denote the transition probabilities from the u_j-th state to the $(j + 1)$-th and the $(j - 1)$-th states, respectively. The state $j = 0$ is the initial neutral state (neither guilty nor high-minded). The states with a positive index are high-minded states, and those with a negative index are guilty states. Being in a high-minded state implies frequent cooperation behavior in the past; being in a guilty state indicates overwhelming defection conduct in the past. The morality factor g_i of u_i is evaluated by a function $f(x_i, st_i)$ that increases as x_i decreases or st_i increases. Later, in Sect. 3.4 when we present our performance evaluation, we will define a specific $f()$.

3.2.2 Design Goal

We address a fundamental tradeoff between the privacy preservation and the data forwarding efficiency in the MSN. Specifically, with the *multiple pseudonym* technique applied for privacy preservation, an unpleasant accompanying side-effect is that users are unable to identify their social friends because of the anonymity of users. This directly impedes the cooperative data forwarding as social ties among users are interrupted. Since users are anonymous, the malicious behaviors (e.g., selfish and free-riding) can no longer be tracked and punished on time using traditional mechanisms.

This may discourage user cooperations and deteriorate the data forwarding efficiency. Therefore, privacy preservation protects and hides the identities of users to the public, which, however, hinders the social-based cooperative data forwarding. Our goal is *to resolve the two conflicting goals in one framework by proposing a privacy preserving social-based cooperative data forwarding protocol*. We exploit the social morality for cooperative data forwarding design. Specifically, the morality of human beings is a common social phenomenon in real-world which provides the rules for people to act upon and grounds the moral imperatives.

3.3 PDF Solutions

3.3.1 Overview of the Protocol

With the user-to-spot data forwarding protocol deployed, in the following sections, we concentrate on how to forward packets to the hotspots for effective and efficient data forwarding with privacy preservation. This delivery is enabled in three steps:

1. Privacy-preserving route-based authentication,
2. Proximity measurement,
3. Morality-driven data forwarding.

In the first step, the privacy-preserving route-based authentication enables two encountered users to exchange partial route information. The route information can be constructed in a privacy-preserving structure determined by users themselves. The use of an authentication scheme is to resist user manipulation attacks, i.e., users have to honestly tell about their hotspots. In the second step, each user measures a proximity score between the destination and the route information provided by the relay user. The proximity score reflects the forwarding capability of a relay node with respect to a specific destination. The larger a proximity score is, the more effective a relay's forwarding is. In addition, the proximity score also affects the morality factor of the relay node. The rationale is that a user would feel more guilty if he/she demonstrates more capability to deliver a packet (with a large proximity score), and yet, drops the packet. In the third step, the morality factor is incorporated into the utility calculation of a data forwarding game in which users act selfishly and preserve their privacy. We elaborate these three steps in the subsequent sections. Note that, we do not consider irrational attacks here. Users tend to be rational and selfish to maximize their own utility.

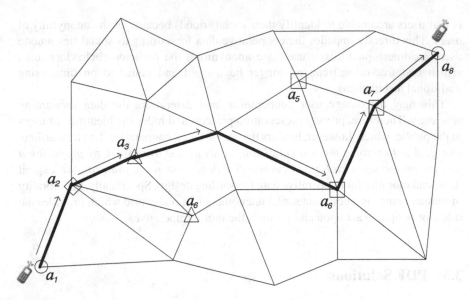

Fig. 3.3 Geographical view of u_i's route

3.3.2   Phase 1: Privacy-Preserving Route-Based Authentication

We first show how to construct a privacy-preserving *routing tree* which describes the route of u_i between hotspots. At an initial stage, the TA associates u_i to a subset of hotspots $\mathscr{A}_i = \{a_y | y = (2,3,6,7)\} \subseteq \mathscr{A}$, which represents the hotspots frequently visited by u_i. We consider that u_i is located at hotspot a_1 and moving towards a_8, as shown in Fig. 3.3. Suppose that u_i moves along the route $a_1 \rightarrow a_2 \rightarrow a_3 \rightarrow a_6 \rightarrow a_7 \rightarrow a_8$. Users neighboring u_i have already known u_i's current location a_1. But u_i has no intention to reveal a_8 to them for privacy reason. In addition, it is unwilling to authenticate the entire hotspot set $\{a_y | y = (2,3,6,7)\}$, which contains privacy-sensitive hotspots $\{a_3, a_6, a_7\}$. Then u_i creates a tree for its mobility route $\mathscr{T}_i$ as "a_2 AND (a_3 OR a_4) AND (2 of (a_5, a_6, a_7))" and only authenticates this tree to others. The authentication reveals the following fuzzy information instead of the precise route: u_i will visit a_2, one of (a_3, a_4), and at least two hotspots from (a_5, a_6, a_7).

We present the routing tree structure $\mathscr{T}$ as shown in Fig. 3.4, where each non-leaf node represents a threshold gate and each leaf node represents a hotspot in $\mathscr{A}_u$. We use $\mathscr{A}_{\mathscr{T}} = \{a_{z_1}, a_{z_2}, \cdots, a_{z_\tau}\} \subseteq \mathscr{A}_u$ to denote the hotspot set corresponding to all leaf nodes in $\mathscr{T}$. Note that, if we assign 0 or 1 to the hotspots $(a_{z_1}, a_{z_2}, \cdots, a_{z_\tau})$ of leaf nodes in $\mathscr{T}$, $\mathscr{T}$ will be transformed into a Boolean function $F(a_{z_1}, a_{z_2}, \cdots, a_{z_\tau})$. For example, in Fig. 3.4, $F(a_1, a_2, \cdots, a_7) = a_2(a_3 + a_4)(a_5a_6 + a_5a_7 + a_6a_7)$. We say that a hotspot set $\mathscr{A}_i$ satisfies both $\mathscr{T}$ and function $F(a_{z_1}, a_{z_2}, \cdots, a_{z_\tau})$ if and only if $F(a_{z_1}, a_{z_2}, \cdots, a_{z_\tau}) = 1$, where for each $a_y, y \in \{z_1, z_2, \cdots, z_\tau\}$,

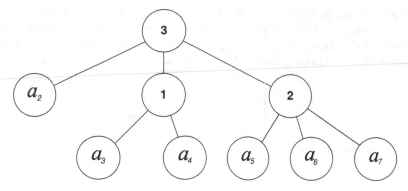

Fig. 3.4 Tree structure of u_i's route

$$a_y = \begin{cases} 1, \text{ if } a_y \in \mathscr{A}_i, \\ 0, \text{ if } a_y \notin \mathscr{A}_i. \end{cases}$$

The routing tree preserves user privacy by making sensitive hotspots anonymous, and at the same time it provides certain information of the mobility route that can be used to evaluate the user's forwarding capability. We are now ready to present our privacy-preserving route-based authentication scheme which supports a single threshold gate (maximum threshold value d) for a routing tree. A multiple-threshold tree can be semantically converted to multiple single-threshold trees. The authentication scheme is built on the bilinear pairing technique [53, 54].

Initialization: Let $\mathbb{G}$ and $\mathbb{G}_T$ be two finite cyclic groups of the same composite order n, where $n = pq$ is a product of two large primes p and q (Sect. 1.4.4.2). Suppose $\mathbb{G}$ and $\mathbb{G}_T$ are equipped with a pairing, i.e., a non-degenerated and efficiently computable bilinear map $e : \mathbb{G} \times \mathbb{G} \to \mathbb{G}_T$ such that (1) $\forall g, h \in \mathbb{G}$, $\forall a, b \in \mathbb{Z}_n, e(g^a, h^b) = e(g, h)^{ab}$; and (2) $\exists g \in \mathbb{G}, e(g, g)$ has order n in $\mathbb{G}_T$.

TA chooses a redundant hotspot set $\mathscr{A}_r = \{a_{l+1}, a_{l+d-1}\}$, two generators (g, u) of $\mathbb{G}$, a generator h of $\mathbb{G}_q$ ($\mathbb{G}_q$ is a subgroup of $\mathbb{G}$ with order q), a secure cryptographic hash function $H : \{0, 1\}^* \to \mathbb{Z}_n^*$, and random number $\delta \in \mathbb{Z}_n^*$. For all $1 \leq y \leq l + d - 1$, TA chooses random numbers $t_y \in \mathbb{Z}_n^*$ and computes $T_y = g^{t_y}$. TA also computes $\Delta = e(g, u)^\delta$. With these settings, TA keeps the master key $(\delta, (t_y)_{1 \leq y \leq l+d-1})$ secretly, and publishes the public parameter $\mathsf{pub} = (n, g, u, h, \mathbb{G}, \mathbb{G}_T, e, H, \Delta, T_y (1 \leq y \leq l + d - 1), \mathscr{A} \cup \mathscr{A}_r)$.

User Registration: TA chooses a unique random number $t \in \mathbb{Z}_n^*$ and a random polynomial $q(x) = \kappa_{d-1} x^{d-1} + \kappa_{d-2} x^{d-2} + \cdots + \kappa_1 x + \delta$, and generates $E_i = \langle k_d, (d_y)_{a_y \in \mathscr{A}_i \cup \mathscr{A}_r} \rangle$, where $k_d = t$ and $d_y = u^{\frac{q(y)}{t+t_y}}$. It informs the registering u_i about the secret key E_i.

Let users u_i and u_j denote the signer and verifier respectively. Denote u_i's routing tree (with a single threshold) by $\mathscr{T}_i$. Let k be the threshold value of the root of $\mathscr{T}_i$ and Θ_i a hotspot set corresponding to $\mathscr{T}_i$'s leaf nodes. $\Phi_i \subseteq \mathscr{A}_i \cap \Theta_i$ is a hotspot set of size k.

Signing by u_i: u_i first chooses a subset $\mathscr{A}_{r'} \subseteq \mathscr{A}_r$ ($|\mathscr{A}_{r'}| = d - k$). Let $\mathscr{A}_{r'}$ be $\{a_{l+1}, \cdots, a_{l+d-k}\}$. Then, for each hotspot $a_y \in \Psi = \Phi_i \cup \mathscr{A}_{r'}$, u_i computes the Lagrange coefficient $\omega_y = \sum_{w|a_w \in \Psi, w \neq y} \frac{0-w}{y-w}$. It randomly selects $r_t, r_p, r_y \in \mathbb{Z}_n^*$ for $a_y \in \Theta_i \cup \mathscr{A}_{r'}$ and computes S_y for $a_y \in \Theta_i \cup \mathscr{A}_{r'}$ as

$$S_y = \begin{cases} d_y^{\omega_y} \cdot h^{r_y}, \text{if } a_y \in \Psi \\ \\ h^{r_y}, \text{if } a_y \in \Theta_i \setminus \Phi_i \end{cases} \tag{3.1}$$

It outputs the signature

$$\sigma_i = \langle \mathscr{T}_i, S_t, S_p, (S_y)_{a_y \in \Theta_i \cup \mathscr{A}_{r'}}, \pi_1, \pi_2 \rangle,$$

where $S_t = g^{k_d} \cdot h^{r_t}$, $S_p = g^{\frac{1}{k_d + H(pid_i)}} \cdot h^{r_p}$,

$$\pi_1 = S_p^{r_t} (g^{H(pid_i)} g^{k_d})^{r_p},$$

and $\pi_2 = \prod_{a_y \in \Psi} (d_y^{\omega_y})^{r_t} \prod_{a_y \in \Theta_i \cup \mathscr{A}_{r'}} (S_t T_y)^{r_y}$.

Verification by u_j: u_j receives σ_i and checks

$$\begin{cases} e(S_t g^{H(pid_i)}, S_p) \overset{?}{=} e(g,g) \cdot e(h, \pi_1) \\ \\ \prod_{a_y \in \Theta_i \cup \mathscr{A}_{r'}} e(S_y, S_t T_y) \overset{?}{=} \Delta \cdot e(h, \pi_2). \end{cases}$$

If the above equations hold, u_j confirms that u_i has pseudonym pid_i and a hotspot set satisfying $\mathscr{T}_i$. The correctness of the verification is from the following mathematical manipulation:

$$e(S_t g^{H(pid_i)}, S_p) = e(g^t h^{r_t} \cdot g^{H(pid_i)}, g^{\frac{1}{t+H(pid_i)}} h^{r_p})$$

$$= e(g,g) \cdot e(h, (g^{\frac{1}{t+H(pid_i)}} h^{r_p})^{r_t} \cdot (g^t g^{H(pid_i)})^{r_p})$$

$$= e(g,g) \cdot e(h, S_p^{r_t} (g^{H(pid_i)} g^t)^{r_p}) = e(g,g) \cdot e(h, \pi_1)$$

$$\prod_{a_y \in \Theta_i \cup \mathscr{A}_{r'}} e(S_y, S_t T_y)$$

$$= \prod_{a_y \in \Psi} e(d_y^{\omega_y}, S_t T_y) \cdot \prod_{a_y \in \Theta_i \cup \mathscr{A}_{r'}} e(h^{r_y}, S_t T_y)$$

$$= \prod_{a_y \in \Psi} e(u^{\frac{\omega_y q(y)}{k_d + t_y}}, g^{k_d} h^{r_t} g^{t_y}) \cdot \prod_{a_y \in \Theta_i \cup \mathscr{A}_{r'}} e(h^{r_y}, S_t T_y)$$

$$= e(g, u)^{\delta} \prod_{a_y \in \Psi} e(u^{\frac{r_t \omega_y q(y)}{k_d + t_y}}, h) \cdot \prod_{a_y \in \Theta_i \cup \mathscr{A}_{r'}} e(h^{r_y}, S_t T_y)$$

$$= \Delta \cdot e(h, \prod_{a_y \in \Psi} (d_y^{\omega_y})^{r_t} \prod_{a_y \in \Theta_i \cup \mathscr{A}_{r'}} (S_t T_y)^{r_y}) = \Delta \cdot e(h, \pi_2)$$

Privacy Discussion: For user privacy preservation, the route-based authentication scheme mixes the hotspot $a_y \in \Psi$ that u_i has with the hotspot $a_y \notin \Psi$ that u_i does not have from Eq. (3.1) by multiplying a subgroup element h. This achieves full-anonymity, i.e., any other user cannot trace the hotspots which are used to generate the signature, because the element h cannot be distinguished from either $\mathbb{G}_p$ or $\mathbb{G}_q$ without p or q known as a priori. The theoretical proof can be found in [53, 54]. Consider that an adversarial user may use the authenticated route information to identify the signer's trace. Without precaution, such misbehavior may violate location privacy. An effective defense mechanism against this privacy violation is to let each user change the routing tree structures of their route information as frequently as the change of their pseudonyms, and also include redundant hotspots into their routing tree. As a result, different users may generate the same routing tree, and the signature cannot be used to link the past/future locations and behaviors of any specific user.

3.3.3 Phase 2: Proximity Measurement

In this section, we develop a novel proximity measurement for implementing the user-to-spot data forwarding protocol. Consider a packet originated from u_j and destined to $\mathscr{D}_j$, which is a hotspot that its intended receiver frequently visits. When u_j meets a u_i, it computes a forwarding score $e_{j,i}$. This score implies u_i's forwarding capability of bringing the packet to $\mathscr{D}_j$. It is subject to multiple factors such as the time-to-live period of the packet, the probability that u_i drops the packet due to limited storage buffer, how close that u_i can be to $\mathscr{D}_j$, when the closest distance will occur, and so on. However, the more factors used, the more personal information revealed, and the less privacy preserved.

To avoid any additional privacy breach, we define that $e_{j,i} = \psi(r_{j,i})$, where $r_{j,i}$ is the smallest distance between $\mathscr{D}_j$ and the hotspots that u_i will visit and $\psi()$ is a monotonically decreasing function of $r_{j,i}$. The smaller $r_{j,i}$, the more closely u_i can deliver the packet to $\mathscr{D}_j$, the larger $e_{j,i}$ by this definition. A particular case is shown in Fig. 3.5. Even if user h appears to move away from $\mathscr{D}_j$, its forwarding, when used, will still be effective since it is going to encounter u_i who will visit

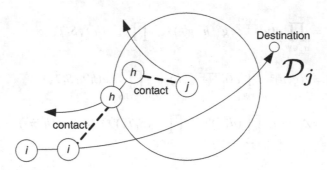

Fig. 3.5 A user moving towards opposite direction of the destination can still provide effective forwarding

$\mathscr{D}_j$ afterwards. Given that no global knowledge is available and any user can be an effective forwarder, $\psi()$ always returns a positive value.

Algorithm 1 Smallest Radius Calculation by u_j

1: **Input:** $\mathscr{T}_i$ and $\mathscr{D}_j$.
2: Transform $\mathscr{T}_i$ to $F_i(a_{z_1}, a_{z_2}, \cdots, a_{z_\tau})$.
3: Calculate $\tilde{F}_i(a_{z_1}, a_{z_2}, \cdots, a_{z_\tau}) = \overline{F_i(\overline{a_{z_1}}, \overline{a_{z_2}}, \cdots, \overline{a_{z_\tau}})}$.
4: Calculate $D_s = \{d_{z_1}, d_{z_2}, \cdots, d_{z_i}\}$, where d_y is the distance between $\mathscr{D}_j$ and a_y for $y \in \{z_1, z_2, \cdots, z_\tau\}$.
5: Sort D_s in an ascending order $\{d_{z_1^*}, d_{z_2^*}, \cdots, d_{z_\tau^*}\}$ corresponding to spots $\{a_{z_1^*}, a_{z_2^*}, \cdots, a_{z_i^*}\}$.
6: Initialize $\tilde{\mathscr{A}} = \{a_{z_1^*}\}, \mu = 1$.
7: **while** ($\tilde{\mathscr{A}}$ does not satisfy $\tilde{F}_i(a_{z_1}, a_{z_2}, \cdots, a_{z_\tau})$) **do**
8:     $\mu = \mu + 1$,
9:     $\tilde{\mathscr{A}} = \tilde{\mathscr{A}} \cup \{a_{z_\mu^*}\}$.
10: **end while**
11: Let $r_{j,i}^* = d_{z_\mu^*}$ and $\mathscr{A}_{\mathscr{D}_j, r_{j,i}^*} = \tilde{\mathscr{A}}$.
12: Output $r_{j,i}^*$.

Since u_i only exposes partial information $\mathscr{T}_i$ of its mobility route to u_j during route-based authentication, u_j cannot compute $r_{j,i}$ accurately. We devise an approximation algorithm for u_j to obtain an approximate value $r_{j,i}^*$ with the inputs $\mathscr{T}_i$ and $\mathscr{D}_j$. In this algorithm, we first transform $\mathscr{T}_i$ to a Boolean function $F_i(a_{z_1}, a_{z_2}, \cdots, a_{z_\tau})$. We denote a self-dual function of F_i as $\tilde{F}_i(a_{z_1}, a_{z_2}, \cdots, a_{z_\tau}) = \overline{F_i(\overline{a_{z_1}}, \overline{a_{z_2}}, \cdots, \overline{a_{z_\tau}})}$. Let $\mathscr{A}_{\mathscr{D}_j, r}$ denote a set of hotspots located in a circular area centered at the destination $\mathscr{D}_j$ with radius r. For a u_i neighboring u_j, we can find the smallest radius $r_{j,i}^*$ such that $\mathscr{A}_{\mathscr{D}_j, r_{j,i}^*}$ satisfies function $\tilde{F}_i(a_{z_1}, a_{z_2}, \cdots, a_{z_i})$. The algorithm finally outputs an approximate value $r_{j,i}^*$. The algorithmic detail is given in Algorithm 1. u_j will then use this value $r_{j,i}^*$ to calculate the forwarding score of u_i.

We use an example to illustrate how proximity score is computed, in accordance with the scenario given in Fig. 3.6. u_i encounters u_j. u_i generates a routing tree

Fig. 3.6 An example of the smallest radius calculation

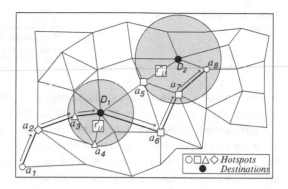

$\mathcal{T}_i$ and the corresponding Boolean function $F_i(a_2, a_3, \cdots, a_7)$ = "a_2 AND (a_3 OR a_4) AND (2 of (a_5, a_6, a_7))". u_j has two packets with destinations $\mathcal{D}_1$ and $\mathcal{D}_2$, respectively. We have $\tilde{F}_i(a_2, a_3, \cdots, a_7)$ = "a_2 OR (a_3 AND a_4) OR (2 of (a_5, a_6, a_7))". According to Algorithm 1, with $\mathcal{T}_i$ and $\mathcal{D}_1$ as inputs, $\tilde{\mathcal{A}}$ is initialized to $\{a_3\}$ since a_3 is the hotspot closest to $\mathcal{D}_1$. Then, $\{a_4\}$ will be added into $\tilde{\mathcal{A}}$ since $\{a_3\}$ does not satisfy $\tilde{F}_i(a_2, a_3, \cdots, a_7)$ and a_4 is the second closest to $\mathcal{D}_1$. $\tilde{\mathcal{A}} = \{a_3, a_4\}$ now satisfies $\tilde{F}_i(a_2, a_3, \cdots, a_7)$. The algorithm finally outputs the distance $r'_{j,i}$ between a_4 and $\mathcal{D}_1$. Similarly, with $\mathcal{T}_i$ and $\mathcal{D}_2$ as inputs, the algorithm outputs the distance $r''_{j,i}$ between a_5 and $\mathcal{D}_2$, where $\tilde{\mathcal{A}} = \{a_5, a_7\}$ satisfying $\mathcal{T}_i$.

3.3.4 Phase 3: Morality-Driven Data Forwarding

After finishing the first two steps, users can perform morality-driven data forwarding. Note that the mobile social users are autonomous and intelligent individuals. It is reasonable to assume that they are rational and their behaviors are driven by personal profit and morality. On one hand, they tend to act defection in order to reduce their forwarding costs. On the other hand, they offer cooperation from time to time so as to counteract the guilty feelings brought by the past selfish deeds. During MSN-based data forwarding, social users implement the best strategy to balance cost and payoff. In this section, we apply game theory to model individual user behavior and obtain the optimal data forwarding strategy.

Consider a scenario where users move along independently and randomly determined mobility routes. Upon the contact with another user, a user would either cooperate or defect for data forwarding. We assume that, for two users that both have packets to send, cooperation is reciprocal. Due to the random mobility and privacy preservation, users' future contacts are unpredictable. A user thus derives the optimal data forwarding strategy based on its self-related information, including its own mobility route, destination of its own packet, and morality factor, as well as the opponent information, including the morality factor, mobility route and packet destination of the encountered user.

Table 3.1 Payoff matrix of B-game

$i \setminus j$	Cooperate (C)	Defect (D)
Cooperate (C)	$(b-c, b-c)$	$(-c, b)$
Defect (D)	$(b, -c)$	$(0, 0)$

From a user's perspective, among a series of cooperations with different encountered opponents, due to the privacy preservation, the opponent information of current contact is always independent from that of previous contacts, and thus the decision on cooperation or defection depends only on the self-related information and the opponent information of the current contact. We thus model the interplay upon each contact, namely cooperation game, as a nonzero sum two-player game.

3.3.4.1 Basic/Extended Cooperation Games

We first define a basic cooperation game, called B-game (B stands for Basic), as a 3-tuple $(\mathcal{N}, \mathcal{S}, \mathcal{P})$, where $\mathcal{N}$ is a pair of users, $\mathcal{S}$ is a set of strategies and $\mathcal{P}$ is a set of payoff functions. According to Sect. 1.4.2, users continuously change their pseudonyms to preserve their privacy. Pseudonym change breaks any relation previously established between two users and as a result they no longer recognize each other. Therefore, B-game is a non-repeated game which can be described as follows:

- *Players*: Two users u_i and u_j belong to the universal user set $\mathcal{V}$. u_j can also be denoted as $-i$. The two users are within the transmission range of each other, and they decide to cooperate or defect, aiming at maximizing their individual payoff.
- *Strategy*: Upon the forwarding request of the opponent user, each user has two strategies: Cooperate (C) and Defect (D). Denote u_i's strategy by s_i. Then $s_i = C$ means that u_i forwards u_j's packet, and $s_i = D$ that u_i drops u_j's packet.
- *Payoffs*: The cost c of forwarding on one packet is a value, the same for both users. If u_i's data is forwarded by u_j, the profit acquired by u_i is b, which is also a constant. We set $b \geq c > 0$ since the profit acquired from each forwarding should be at least equal to the incurred cost. The user payoffs under different strategies are shown in Table 3.1.

From the payoff matrix in Table 3.1, it is observed that the B-game is a typical prisoner-dilemma game, where the only Nash Equilibrium (NE) is (D, D) for non-repeated version. In other words, no matter what the opponent's strategy is, the best strategy for a user is to defect. This is because that $b > b - c, 0 > -c$.

Next, we introduce an E-game (E stands for Extended), where the payoff matrix is shown in Table 3.2. This game considers users u_i and u_j's behaviors affected by morality factors g_i and g_j. The morality factors are introduced as the costs of defection behaviors into the payoff functions. The best strategy of the E-game for u_i is: cooperate if $g_i > c$; defect if $g_i \leq c$. Based on the Markov chain model

Table 3.2 Payoff matrix of
E-game

$i \setminus j$	Cooperate C	Defect D
C	$(b - c, b - c)$	$(-c, b - g_j)$
D	$(b - g_i, -c)$	$(-g_i, -g_j)$

given in Sect. 3.2.1, there exists a morality state $x^* < 0$ such that $f(st_i, x^* + 1) < c < f(st_i, x^*)$. After a finite series of defections, u_i will reach state x^*, and then alternatively chooses to cooperate.

3.3.4.2 Social Cooperation Game

In the following, we extend the E-game to a complex S-game (S stands for Social), which is also denoted by a 3-tuple $(\mathcal{N}, \mathcal{S}, \mathcal{P})$. S-game further incorporates the forwarding scores $e_{i,j}$ and $e_{j,i}$ into the payoff function.

- *Players*: Two users u_i and u_j with different sociality strength st_i, st_j and current morality factors g_i, g_j.
- *Strategy*: The strategy is the same as that of the B-game. u_i's strategy is denoted by s_i.
- *Payoffs*: The payoff of u_i is evaluated by

$$
p_i^s = \begin{cases} e_{i,j}b - c, & \text{if } s_i = C, s_j = C, \\ -c, & \text{if } s_i = C, s_j = D, \\ e_{i,j}b - e_{j,i}g_i, & \text{if } s_i = D, s_j = C, \\ -g_i, & \text{if } s_i = D, s_j = D. \end{cases} \tag{3.2}
$$

In payoff formula (3.2), the forwarding scores $e_{i,j}$ and $e_{j,i}$ are used to measure u_i's profit and morality factor. If u_j forwards u_i's data, the profit that u_i acquires is $e_{i,j}b$ instead of b. If u_i drops u_j's data, depending on u_j's strategy, u_i acquires different morality factors, $e_{j,i}g_i$ or g_i. Note that, when users u_i and u_j both drop each other's packets, the morality factor on u_i's payoff is independent of the forwarding score $e_{j,i}$. This is because users u_i and u_j treat each other equally and do not further consider their forwarding capability.

3.3.4.3 S-Game with Complete Information

We first analyze the S-game in the case that two players have complete information including the sociality strength and morality state of each other. Each player can calculate the morality factor by $\psi()$ as defined in Sect. 3.2.1 and determine the payoff before deciding whether to cooperate or defect, according to Table 3.3. We use Theorem 3.1 to identify the NE strategies of the S-game.

Table 3.3 Payoff matrix of S-game

$i \setminus j$	Cooperate C	Defect D
C	$(e_{i,j}b - c, e_{j,i}b - c)$	$(-c, e_{j,i}b - e_{i,j}g_j)$
D	$(e_{i,j}b - e_{j,i}g_i, -c)$	$(-g_i, -g_j)$

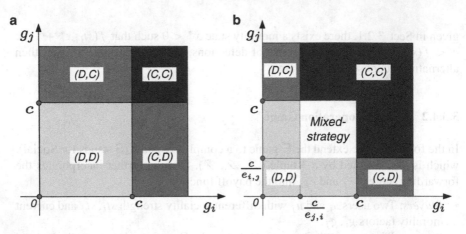

Fig. 3.7 The best strategy for different games. (**a**) The best strategy of E-game. (**b**) The best strategy of S-game

Theorem 3.1. *When the two players have complete information of each other in the S-game, there are multiple pure-strategy NE in different cases and one mixed-strategy NE (x_i, x_j), where $x_i = \frac{c - g_{-\theta}}{e_{\theta,-\theta}g_{-\theta} - g_{-\theta}}$ is the probability that user n_θ chooses to cooperate, as shown in Fig. 3.7b.*

Proof. For $\theta = i$ or j, we have the following three cases to consider.

- $g_\theta < \frac{c}{e_{-\theta,\theta}}$: We have $e_{\theta,-\theta}b - e_{-\theta,\theta}g_\theta > e_{\theta,-\theta}b - c$ and $-g_\theta > -\frac{c}{e_{-\theta,\theta}} \geq -c$ due to $e_{-\theta,\theta} \geq 1$. As a result, when $g_{-\theta} < c$, $(s_\theta = D, s_{-\theta} = D)$ is a NE; and when $g_{-\theta} > c$, $(s_\theta = D, s_{-\theta} = C)$ is a NE.

- $g_\theta > c$: We have $-g_\theta < -c$ and $e_{\theta,-\theta}b - e_{-\theta,\theta}g_\theta < e_{\theta,-\theta}b - c$ due to $e_{-\theta,\theta} \geq 1$. As a result, when $g_{-\theta} > \frac{c}{e_{\theta,-\theta}}$, $(s_\theta = C, s_{-\theta} = C)$ is a NE; when $g_{-\theta} < \frac{c}{e_{\theta,-\theta}}$, $(s_\theta = C, s_{-\theta} = D)$ is a NE.

- $\frac{c}{e_{-\theta,\theta}} < g_\theta < c$: Let x_θ denote the forwarding probability of user n_θ. For $s_{-\theta} = C$ or $s_{-\theta} = D$, we separately calculate the payoff for $n_{-\theta}$ as follows:

$$p^s_{-\theta}|C = x_\theta \times (e_{-\theta,\theta}b - c) + (1 - x_\theta) \times (-c)$$

$$p^s_{-\theta}|D = x_\theta \times (e_{-\theta,\theta}b - e_{-\theta,-\theta}g_{-\theta}) + (1 - x_\theta) \times (-g_{-\theta})$$

If x_θ is the best strategy of n_θ, we have $p^s_{-\theta}|C = p^s_{-\theta}|D$ which gives $x_\theta = \frac{c - g_{-\theta}}{e_{\theta,-\theta}g_{-\theta} - g_{-\theta}}$.

$\square$

3.3.4.4 S-Game with Incomplete Information

We consider the case that the two players have incomplete information of each other. Specifically, u_i obtains sociality strength st_i, morality state g_i, forwarding scores $e_{i,j}$ and $e_{j,i}$, but it does not obtain the sociality strength st_j and morality factor g_j of u_j. As a supplementary information, we assume that u_i obtains the probability distribution ϱ of the morality factor of all users. Based on this, u_i can estimate the morality factor g_j of u_j. Then, u_i follows the following steps according to the best strategy shown in Fig. 3.7b:

- If $0 \le g_i < \frac{c}{e_{j,i}}$, then u_i chooses to defect regardless of u_j's strategy.
- If $c \le g_i$, then u_i chooses to cooperate regardless of u_j's strategy.
- If $\frac{c}{e_{j,i}} \le g_i < c$, then there exists a pure-strategy NE (D, D) for $g_j < \frac{c}{e_{i,j}}$, a pure-strategy NE (C, C) for $g_j > c$, and a mixed-strategy NE for $\frac{c}{e_{i,j}} < g_j < c$. For the pure strategy NE, we calculate the defection probability Pr_1 and cooperation probability Pr_2:

$$Pr_1 = Pr(0 \le g_j < \frac{c}{e_{i,j}}) = \int_0^{\frac{c}{e_{i,j}}} \varrho(\alpha)d\alpha,$$

$$Pr_2 = Pr(c \le g_j) = \int_c^{+\infty} \varrho(\alpha)d\alpha.$$

In addition, u_i makes a mixed-strategy NE with probability Pr_3, which is given by

$$Pr_3 = Pr(\frac{c}{e_{j,i}} \le g_i < c) = \int_{\frac{c}{e_{j,i}}}^c \varrho(\alpha)d\alpha.$$

For the mixed-strategy NE with probability Pr_3, Theorem 3.1 indicates the best strategy of u_i is to forward the data with probability $\frac{c-g_j}{e_{i,j}g_j-g_j}$ if g_j is known by u_i. In this case, the probability that u_i chooses to cooperate is

$$Pr_4 = \int_{\frac{c}{e_{j,i}}}^c (\frac{c-\alpha}{e_{i,j}\alpha - \alpha})\varrho(\alpha)d\alpha. \tag{3.3}$$

Overall, u_i decides to cooperate with probability $Pr_F = Pr_2 + Pr_4$ and to defect with probability $Pr_D = 1 - Pr_F$.

3.3.5 Summary of Data Forwarding Strategy

Notice that the S-game with incomplete information emulates MSN environments in reality, where the opponent's morality factor cannot be directly obtained. We use the optimal strategy of this game in our protocol for users to make the optimal data forwarding strategies. As we defined in Sect. 3.2.1, user morality factor would vary with both sociality strength and morality state. However, revealing such information violates user privacy since other adversarial users can utilize the information to track user behavior. In this case, we do not require an accurate calculation of morality factor in the S-game. Instead, we examine the strategy by using a probability distribution function ϱ of morality factor. This function ϱ can be either observed by a trusted authority or reported by individual users. Further analysis is presented in Sect. 3.3.4.4.

A user who has packets to forward starts the data forwarding protocol with a randomly selected neighbor. Consider two neighboring users u_i and u_j that are running the protocol, i.e., they are both able to provide cooperative data forwarding to each other and any forwarding/defection decision in the two-user game will impact their social morality. Let $S_i = \{p_{i_1}, p_{i_2}, \cdots, p_{i_\alpha}\}$ and $S_j = \{p_{j_1}, p_{j_2}, \cdots, p_{j_\beta}\}$ be the packet sets held by u_i and u_j, respectively. We summarize the protocol as follows. u_i first randomly selects a packet p_x (destined to $\mathscr{D}_i$) from its local repository. It then calculates the digest of the packet $d_i = H(p_x)$, where H is the cryptographic hash function. Lastly, it sends d_i to u_j. In the meantime, u_j executes a similar procedure locally and sends u_i the digest d_j of a packet of p_y (destined to $\mathscr{D}_j$). According to d_i (d_j), if u_j (resp., u_i) finds that it already has p_x (resp., p_y), it will inform u_i (resp., u_j) to re-select p_x (resp., p_y). Through exhaustive packet re-selection, if they cannot find any exclusively owned packet, the protocol will terminate. Otherwise, they proceed to exchange $(p_x, \mathscr{D}_i)$ and $(p_y, \mathscr{D}_j)$, together with their own routing trees $\mathscr{T}_i$ and $\mathscr{T}_j$. Then, they validate each other's routing trees (see Sect. 3.3.2). After that, they evaluate each other's forwarding scores (see Sect. 3.3.3), and finally make the forwarding strategy for each other (see Sect. 3.3.4.4).

3.4   Performance Evaluation

3.4.1   Simulation Settings

3.4.1.1   User Mobility, Hotspots, and Packet Generation

We generate user mobility model according to the real-world trace of pedestrian runners provided in [77]. In the real trace set, $N = 100$ mobile users are randomly deployed in a $1,000 \times 1,000 \, \text{m}^2$ square region with the velocity randomly distributed with a mean value of 1 m/s. The communication range R_t of users is set to 50 m. The log contains the user locations in successive $T = 900$ time slots.

We divide the network field into 10×10 grids, where each grid is a square with side length 100 m. We create a circle of radius R_t around each grid point, and there are totally 121 circles. The areas enclosed by these circles are called spots and denoted by $(a_1, a_2, \cdots, a_{121})$ as shown in Fig. 3.8; no any two spots overlap.

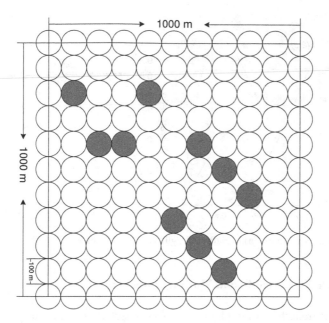

Fig. 3.8 Hotspots

We aggregate user route information to determine the most popular spots as follows. Let $d_{m,n}$ denote the number of users in hotspot a_m at time slot n, where integers $m \in [1, 121]$ and $n \in [1, 900]$. We sort the spots in an descending order according to $d_m = \sum_{n=1}^{T} d_{m,n}$, and choose the top-ten spots as hotspots (i.e., $l = 10$). At the middle of each hotspot, we place a wireless storage device which has a communication range equal to R_t. Once a user enters a hotspot, it can access the storage device of the hotspot via wireless communication.

For each simulation run, there are totally 1,000 packets generated for transmissions, 100 packets per each user, with uniformly selected hotspots as the packet destinations. In each time slot, a u_i randomly selects a neighboring u_j to play a two-player cooperation game. In the cooperation game, we consider the communication cost of data forwarding to be much greater than the computational cost of the associated authentication. As such, the authentication scheme imposes negligible influence on user behavior. Upon each contact, users uniformly select one available packet from their buffers to transmit. In order to focus on the impact of cooperation on the data forwarding effectiveness, we consider packets do not expire during the simulations and hotspot buffers and user device buffers are not limited in size.

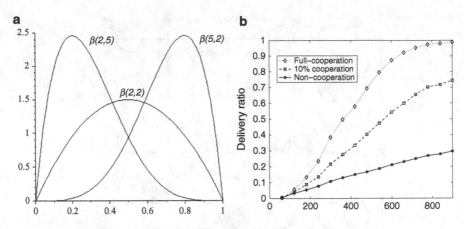

Fig. 3.9 Preliminary results. (**a**) Sociality strength. (**b**) Cooperation effect

3.4.1.2 Sociality Strength and Morality Function

The sociality strength st_i of u_i ($1 \leq i \leq 100$) is selected from the range of $[0, 1]$. The greater st_i is, the more intense social morality impact on u_i's cooperation. In this section, we adopt different models of sociality strength represented by three beta distributions $\beta(2, 5)$, $\beta(2, 2)$, $\beta(5, 2)$ shown in Fig. 3.9a, respectively, to evaluate the performance of the protocol in the cases of low, medium and high users' sociality strength, respectively.

The morality function f is used to calculate the morality factor of each u_i using the user's sociality strength st_i and current morality state x. From Sect. 3.2.1, we define three morality functions: *linear* function f_1, *natural logarithm* function f_e and *common logarithm* function f_{10}. They output 0 if $x \geq 0$, and otherwise,

$$f_1(st_i, x) = k \cdot st_i \cdot (-x)$$

$$f_e(st_i, x) = k \cdot \ln(1 + st_i \cdot (-x))$$

$$f_{10}(st_i, x) = k \cdot \log_{10}(1 + st_i \cdot (-x))$$

where k is a tunable coefficient in the range of $(0, +\infty)$. For simplicity, we fix $k = 1$ in our simulation.

The three morality functions represent three different levels of morality force affecting user cooperation behavior, respectively. They always output a non-negative value. The *common logarithm* function f_{10} generates a smaller morality factor, compared with the other two functions. If it is adopted, we can expect to see more defection behaviors.

3.4.1.3 Routing Tree and Forwarding Capability

Recall that a user's routing tree preserves user privacy by making the sensitive hotspots anonymous, and in the meantime provides partial information of user mobility route in order to facilitate cooperative data forwarding. With 10 hotspots in simulations, each u_i may have at most 10 hotspots and at least 0 hotspot in $\mathscr{A}_i$. We generate a simplified routing tree structure $\mathscr{T}$ in the following way: if $|\mathscr{A}_i| = 0$, the tree cannot be created; if $0 < |\mathscr{A}_i| < 5$, we set the threshold as $|\mathscr{A}_i|$, and the leaf nodes as all the hotspots of $\mathscr{A}_i$ and other $5 - |\mathscr{A}_i|$ ones from $\mathscr{A}_u \setminus \mathscr{A}_i$; if $|\mathscr{A}_i| \geq 5$, we set the threshold as 4, and the leaf nodes as four randomly selected hotspots from $\mathscr{A}_i$ and another different hotspot. In short, for every user, the tree structure can be written as "t of 5", where $1 \leq t \leq 4$.

In Sect. 3.3.3, a function ψ is used to compute the forwarding capability of a given u_i for a packet with a specific destination. We set the lower bound of ψ as 1. In the network grid, $r_{i,j}^*$ can be $1,000 \times \sqrt{2} = 1415$ m at most and 0 at least. Intuitively, if $r_{i,j}^* = 1,415$, the forwarding capability $e_{i,j}$ reaches the minimum value; and if $r_{i,j}^* = 0$, $e_{i,j}$ reaches the maximum value. We define $\psi(r_{i,j}^*) = e^{k'-k'r_{i,j}^*/1415}$ and set $k' = 3$ as an example to illustrate the effect of forwarding capability.

3.4.2 Simulation Results

The performance metrics used in the simulation are: (1) the delivery ratio, which is the fraction of packets that are correctly delivered to the hotspots as their destinations; and (2) the average morality state, which reflects the intention of users to cooperate over time. The delivery ratio examines the overall cooperation of users in the MSN, while the average morality state denotes the long-term cooperation strategies for a single user. For each simulation, we conduct 50 simulation runs and report the average results.

3.4.2.1 B-Game

We first examine the B-game, where users always choose defection as the best strategy as discussed in Sect. 3.3.4.1. Figure 3.9b shows three delivery ratios in the following three cases: (a) users do not cooperate (i.e., B-game); (b) users stochastically cooperate to forward packet with the probability of 10 %; and (c) users fully cooperate. It can be seen that at time slot 900, the full-cooperation strategy achieves 99 % delivery ratio while the non-cooperation strategy achieves only 30%. Furthermore, Fig. 3.9b indicates that the probabilistic cooperation strategy provides a significant improvement to the delivery ratio up to 74 %. However, without effective incentive and appropriate exploration of their social feature, users will not take cooperation due to the selfishness. Successful delivery happens only when the data senders arrive at their selected hotspots. This inevitably results in a low delivery ratio in the B-game.

3.4.2.2 E-Game and S-Game

The E-game extends the B-game by embedding the morality factor into the payoff function as shown in Table 3.2, while the S-game further considers the forwarding capability into the payoff of the E-game. Figure 3.10 shows the delivery ratio of both the E-game and the S-game with complete information, with red lines representing the performance of forwarding cost $c = 0.5$, blue lines representing that of $c = 1.5$, and black lines depicting those of full-cooperation and non-cooperation as the best and worst case. It is clearly observed that the strategies with $c = 0.5$ can achieve higher delivery ratio than the strategies with $c = 1.5$. The rationale is that a large forwarding cost $c = 1.5$ hinders the cooperation performed by users who have limited resources and thus limits guilty incentive. In particular, when f_{10} is adopted, the cooperation condition in case of $c = 1.5$ approaches to the worst case. This is because that the guilty function f_{10} returns the smallest morality factor resulting in the least incentive to cooperate, compared to the function f_e and f_1. Figure 3.10 shows that the strategies with the sociality strength $\beta(5, 2)$ perform much better than those with $\beta(2, 2)$ and $\beta(2, 5)$ in terms of delivery ratio. This is because that, compared to cases $\beta(2, 2)$ and $\beta(2, 5)$, users will be initialized with larger sociality strength in case $\beta(5, 2)$ as shown in Fig. 3.9a, and as discussed in Sect. 3.2.1, more users feel intense guilt towards their defection and choose to cooperate, which leads to a better performance.

Figure 3.10 shows the performance comparisons between the E-game and the S-game under the same parameters. It can be seen that the delivery ratio can be further improved by enabling privacy-preserving route-based authentication. But since the route information is limited due to the privacy preservation, the improvements are not significant, e.g., when choosing $\beta(2, 5)$ and $c = 1.5$, the delivery ratio increases from 0.309 as shown in Fig. 3.10c to 0.327 as shown in Fig. 3.10f. To further investigate the impact of the route information on the data forwarding cooperation, we randomly select 100 users in the network and examine their average morality states. Figure 3.11 shows the average morality state of each selected user in terms of the user sociality strength in three settings of social strength $\beta(2, 5)$, $\beta(2, 2)$, and $\beta(5, 2)$, respectively. The blue circle represents a user which adopts the best strategy from the S-game, and the red star represents a user which adopts the best strategy from the E-game. It can be seen that with the same sociality strength, the users represented by the red star have smaller morality states than users represented by the blue circle. This is to say, the incentive to defect in the cooperation game can be further reduced by enabling privacy-preserving route-based authentication.

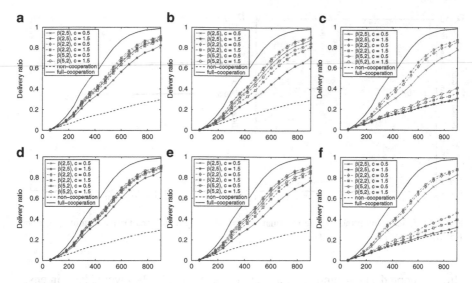

Fig. 3.10 Delivery ratio in E-game and S-game with complete information. (**a**) E-game with f_1. (**b**) E-game with f_e. (**c**) E-game with f_{10}. (**d**) S-game with f_1. (**e**) S-game with f_e. (**f**) S-game with f_{10}

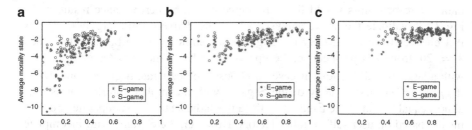

Fig. 3.11 Average morality states of all users in E-game and S-game with complete information, $c = 0.8$, and common logarithm. (**a**) Sociality strength $\beta(2, 5)$. (**b**) Sociality strength $\beta(2, 2)$. (**c**) Sociality strength $\beta(5, 2)$

3.4.2.3 S-Game with Incomplete Information

For the S-game with incomplete information, the morality factor cannot be obtained directly in our morality model due to the lack of sociality strength and morality state information about the opponent user. As such, the morality factor will be estimated by a probability distribution function. In our simulation, we use exponential distribution with parameter $\lambda = \{1, 2, 10, 20\}$ to generate the morality factors for all users. The probability distribution function is shown in Fig. 3.12a.

Figure 3.12a shows that most users in case of $\lambda = 1$ may have relatively large morality factor. As we make $st_i \sim \beta(2, 5)$, most users would have the weak sociality strength. Thus, the large morality factors of users indicate that they have

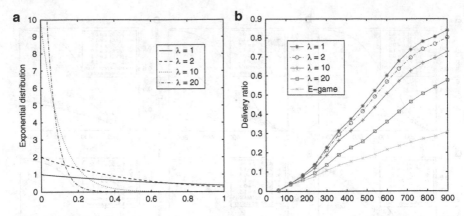

Fig. 3.12 S-game with incomplete information. (**a**) Morality factor. (**b**) Delivery ratio, f_{10}, $c = 1.5$, $st_i \sim \beta(2, 5)$

already adopted a large amount of defections. Accordingly, they would have intense guilty feeling so that their following behaviors are probably cooperative. Besides that, it can be seen that when $\lambda = 20$ most users with the weak sociality strength have smaller morality factors, and without enough guilt as cooperation incentives their future behaviors would likely be defections. The performance results from Fig. 3.12b validate the above analysis, where the delivery ratio largely decreases if λ changes from 1 to 20. By investigating the strategy, it can be seen that when $\lambda = 20$, from u_i's perspective, the opponent u_j has a morality factor $g_j < \frac{c}{e_{i,j}}$ with a large probability. In this case, u_i chooses to cooperate if $g_i \geq c$ and defect if $g_i < c$. The best strategy of the S-game with incomplete information is thus almost equal to that of the E-game; both games indicate users to cooperate or defect mostly based on user self morality factors. However, the S-game with incomplete information outperforms the E-game since it has an additional mixed-strategy space shown in Fig. 3.7b to encourage user cooperation.

3.5   Related Work

Data forwarding protocols have been extensively investigated in delay-tolerant networks. Due to the sparse and dynamic nature of delay-tolerant networks, user-to-user data forwarding often relies on the mobility and random contacts of users. For example, Lindgren et al. [101] evaluated the forwarding capability of a user by the historic contact information. Under the similar framework, Refs. [16, 45, 98, 102, 103] used social metrics calculated from the contact information to evaluate the forwarding capability of a user. Hui et al. [103] demonstrated that community and centrality social metrics can be effectively used in data forwarding protocol. Li et al. [16] introduced the social-selfish effect into user behavior, i.e., a user gives preference to packets received from other users with stronger social relations.

Yuan et al. [98] proposed a data forwarding protocol enabling two users to share their historical mobility information. Based on the opponent's past mobility information, a user is able to predict the future location that the opponent will visit.

Though significantly improving the data forwarding effectiveness, most contact-based data forwarding protocols require a contact calculation phase in which each user must have a unique identity and reveal it to others. In this phase, user behaviors are very easy to be linked together and user's identity privacy and location privacy are completely compromised. In the contact-based data forwarding protocol, a sender must exchange the contact and unique identity with a relay user. In [16, 45, 101], to improve the forwarding effectiveness, a sender can evaluate the forwarding capability of a relay user based on both the relay user's contact probability and forwarding willingness. However, the required contact probability and unique identity information are privacy-sensitive to the relay user and not available in a privacy-preserving environment. The conventional contact-based data forwarding protocols do not provide effective privacy preservation and can hardly be accepted by the privacy-aware mobile users. We aim to solve the privacy preservation and security issues of cooperative data forwarding in the MSN.

Recently a rich body of literature [104–109] addressed the cooperation stimulation issue from a game-theoretic perspective. Yu and Liu [104] proposed a game-based approach to stimulate cooperation in mobile ad hoc networks, where two participating users set a threshold on the number of forwarded packets in each forwarding round and they alternatively forward each other's packets. The setting of the threshold can stimulate cooperation and also limit the possible damage caused by the opponent's defection. If the opponent defects, a user immediately terminates the interaction and his maximum damage is bounded by the threshold setting in the previous round. Li and Shen [109] proposed an integrated system over an individual reputation system and a price-based system which demonstrates a superiority in terms of the effectiveness of cooperation incentives and selfish node detection. However, their works do not address user privacy and are not applicable in the privacy-sensitive MSN.

The studies in the MSN mainly focus on exploring the human factors and behaviors for communications in a distributed and autonomous environment. Privacy preservation as a fundamental user requirement is however neglected in previous research. Recent proposals [110] indicated that one or a few snapshots of a user's location over time might assist an adversary to identify the user's trace, and an effective attack was presented to identify victims with high probability. As a defense technique, the *multiple-pseudonym* technique providing both identity and location privacy is widely applied in literatures [75, 111, 112]. Freudiger et al. [75] developed a user-centric location privacy model to measure the evolution of location privacy over time, and they derive the equilibrium strategies on changing pseudonyms for each user from the game-theoretic perspective. With the *multiple-pseudonym* technique applied, conventional cooperation stimulation mechanisms without privacy preservation [71, 104, 113, 114] are no longer applicable in the considered environment.

3.6 Conclusion and Future Directions

We have studied two fundamental problems in the MSN, i.e., privacy preservation and cooperative data forwarding. We have indicated the difficulties to solve both problems at the same time. This is because that concealing and protecting user information may prohibit tracking the social behavior of users, which impedes the cooperative data forwarding and effective incentive mechanism. We have attained the two conflicting design goals in one framework. Specifically, we have introduced a novel user-to-spot data forwarding protocol where each packet is destined to a hotspot associated with the receiver and then retrieved by the receiver upon its access to the hotspot. With this protocol, not only can receiver location privacy be preserved, but the packet delivery ratio is also enhanced. Game-theoretic approaches have been adopted to derive the best data forwarding strategy for users, with respect to user morality factor and forwarding capability. Through extensive trace-based simulations, we have demonstrated the data forwarding effectiveness of the protocol in terms of packet delivery ratio. Particularly, the embedded privacy-preserving route-based authentication scheme makes important contribution to the protocol performance.

When we consider user social behavior, we could find many interesting phenomenon. For example, users may prefer to help others that they recognize or trust with; users may rely more on others who have more similarities or belong to the same organizations. In reality, most users may follow some implicit but common patterns. These patterns are very difficult to be extracted and formally defined. However, these patterns influence users' decisions in every aspect including data forwarding strategy. The existing open problems for cooperative data forwarding in the MSN are

- Explore diverse behavior models of different users. In this chapter, we use a Markov-chain to model the morality factor of users. However, the model still has some deficiency where it cannot explain some extreme case of user behavior. For example, when a user takes a defection behavior, he may immediately feel guilty and behave cooperative in the next time slot. But his cooperative incentive may decrease continuingly as memory fades and time passes. The behavior model by the Markov-chain cannot describe such effect. As such, one of the important future works is to further explore the social theory and complete the behavior model design.
- In this chapter, we have introduced a user-to-spot strategy where the location information can be used to evaluate the forwarding capability and the location information can be used as a destination of the receiver. There is another category of data forwarding study that focuses on user-to-user strategy. In most of the user-to-user strategies, the identity information is often required to be disclosed such that users could choose a better relay to deliver the packets. However, most user-to-user strategies do not consider identity privacy of users. How to protect the identities and still achieve effective user-to-user data forwarding is a significant research direction.

Chapter 4
Recommendation-Based Trustworthy Service Evaluation

4.1 Introduction

In this chapter, we introduce another research topic, called trustworthy service evaluation (TSE). We envision the MSN as business streets as shown in Fig. 4.1 where service providers (vendors) are densely distributed. Such MSN is also called Service-oriented MSN (S-MSN). Users in the S-MSN not only want to talk to other users, but also expect to well communicate with the nearby vendors. In the meantime, the vendors will try possible advertising methods to attract the potential customers.

The TSE is a distributed system involving both users and vendors. The intuition of designing the TSE is from the successful business solutions, as shown in Fig. 4.2. When we visit the business stores, especially restaurants, we often see some photos on the walls showing the famous people had a very pleasure time with the restaurant owners. This is a simple but very effective recommendation mechanism; people believe they would have the same experience as the famous people had. The behavior of the famous people with positive reputation would largely impact the customers' choices. In the TSE, we aim to change the format of the recommendations, i.e., convert the photos to non-forgeable digital review comments. The review comments can be collected and disseminated not only inside of the stores but also over the streets. However, the efficiency, security and privacy issues would be more challenging.

We introduce a basic trustworthy service evaluation (bTSE) system and an extended Sybil-resisted TSE (SrTSE) system for the S-MSN. In both systems, no third trusted authorities are involved, and the vendor locally maintains reviews left by the users. The vendor initializes a number of tokens, which are then circulated among the users to synchronize their review submission behaviors. After being serviced by a vendor, a user generates and submits a non-forgeable review to the vendor. The user cannot proceed with the review submission until it receives a token from the vendor. If the review submission succeeds, the user will forward the token to a nearby user who is wishing to submit a review to the same vendor; otherwise,

Fig. 4.1 Mobile social network with vendors

Fig. 4.2 Restaurants with the reviews from famous people

the user will forward both the token and its own review to the receiver, expecting that receiver–user will cooperate and submit their reviews together. During token circulation, a hierarchical signature technique [53, 54] is adopted to specify and record each forwarding step in the token, and a modified aggregate signature technique [80] is employed to reduce token size. Both signature techniques are also used during cooperative review submission for reducing communication overhead and improving review integrity. Specifically, we identify three unique review attacks, i.e., review linkability attack, review rejection attack, and review modification attack in the bTSE. We also introduce two typical sybil attacks which cause huge damage to the bTSE. Under the sybil attacks, the bTSE system cannot work as expected because a single user can abuse the pseudonyms to generate multiple unlinkable false reviews in a short time. To resist such attacks, in the SrTSE, the pseudonyms are embedded with a trapdoor; if any user leaves multiple false reviews toward a vendor in a pre-defined time slot, its real identity will be revealed to the public. Through the security analysis and numerical results, we show that both the bTSE and the extended SrTSE are secure against the possible attacks. We further evaluate the performance of the bTSE in comparison with a non-cooperative system that does not engage cooperative review submission. Simulation results indicate that the bTSE achieves significantly (up to 100%) higher submission rates in the presence of the review rejection attacks, and (up to 75%) lower submission delays in general than the non-cooperative system, at the cost of reasonable cooperation overhead.

The remainder of this chapter is organized as follows: in Sect. 4.2, we present the system model and design goal. Then, we introduce the TSE systems in Sect. 4.3 where the above challenges can be resolved. We provide the security analysis and the simulation-based performance evaluation in Sects. 4.4 and 4.5, respectively. We also review the related works in Sect. 4.6. Lastly, we draw our summary in Sect. 4.7.

4.2 System Model and Design Goal

4.2.1 System Model

As mentioned in previous chapter, an S-MSN may contain multiple vendors offering different or similar services to users. Without loss of generality we consider a singe-vendor S-MSN. There is no central trusted authority in the network. The vendor is assumed to offer a single service. The vendor is equipped with a wireless communication device that has a large storage space. Each user has a handheld device such as cell phone; the transmission range of the device is the same for all users, and smaller than the vendor's transmission range. There are two types of communication in the network: vendor-to-user communication and user-to-vendor communication. The former may take place directly if users are in the vendor's transmission range, or indirectly through other users' message relay otherwise. It aims to disseminate up-to-date service information including service description and reviews to users. The later enables users to submit reviews to the vendor. Similar to vendor-to-user communication, it occurs directly if the vendor is in the transmission range of users, or indirectly otherwise.

We consider a homogenous S-MSN composed of a set $\mathcal{V} = \{u_1, \cdots, u_N\}$ of mobile users with the network size $|\mathcal{V}| = N$. Users have equal communication range, denoted by R_t. From a social perspective [115], users spontaneously form different social groups based on their common interests, termed as "attributes". Suppose that there are p social groups $\{g_1, \cdots, g_p\}$. Let $\mathcal{A}_u$ be the universal attribute set. Denote a social group g_h's attribute set by $\mathcal{A}_h$ ($\mathcal{A}_h \subseteq \mathcal{A}_u$) for $1 \leq h \leq p$. Every user u_j belongs to at least one social group. It inherits the attributes of the social groups that it belongs to. Thus, the attribute set of u_j is $\mathcal{P}_j = \bigcup_{h \in \mathcal{H}} \mathcal{A}_h$, where u_j is a member of g_h. The vendor (precisely, its service) is also tagged by an attribute set $\mathcal{V} \subseteq \mathcal{A}_u$. It periodically disseminates its up-to-date service information including service description and reviews to users. The integrity and non-forgeability of such service information will be achieved by using a public/private key pair of the vendor.

Each group g_h relies on a group authority c_h for membership management. c_h has a public/private key pair (pk_h, sk_h), and publishes the public key to all users. A multi-authority identity based signature scheme [80] is used to implement group membership. Note that c_h is not a part of the network, and the management of group membership is performed offline. Every user u_j has a private unique identity id_j. When it joins g_h, c_h verifies the validity of u_j's identity id_j and assigns u_j a number of randomly generated verifiable pseudonyms $pid_{j,h,1}, pid_{j,h,2}, \cdots$. It also allocates u_j a number of secret keys $psk_{j,h,*}$, each corresponding to $pid_{j,h,*}$. Thus, u_j has a set of pseudonyms $pid_{j,*,*}$ allocated by the group authorities of the social groups that it belongs to. It interacts with the vendor and other users using these pseudonyms alternatively, instead of its unique identity id_j, for privacy preservation. Reviews are associated with pseudonyms, which in turn belong to social groups, such that the vendor and other users are able to check the authenticity of the reviews and the group authorities are able to trace the reviews generated by their group members.

4.2.2 Design Goal

Due to the lack of centralized control, the S-MSN is vulnerable to various security threats. It is worthy noting that with third trusted authorities in the network, the security problems can be easily solved. We consider that the group authorities are trusted but not a part of the network. The vendor and compromised users can manipulate reviews for malicious competition. In the following, we describe several malicious attacks that aim particularly at the TSE. They are called *review attacks* and *sybil attacks*, where the review attacks includes review linkability, rejection and modification attacks and the sybil attacks have two categories. Without protection, they may take place easily, paralyzing the entire system.

Review Attack 1: Review linkability attack is executed by malicious users, who claim to be members of a specific group, but disables the group authority to trace the review back to its unique identity, thus breaking review linkability.

Review Attack 2: Review rejection attack is launched by the vendor when a user submits a negative review to it. In the attack, the vendor drops the review silently without responding to the submission request from the user. The vendor may intend to perform review rejection attacks so as to hide public opinions and mislead users.

Review Attack 3: Review modification attack is performed by the vendor toward locally recorded review collections. The vendor inserts forged complimentary reviews, or modifies/deletes negative reviews in a review collection. Such attacks aim at false advertising by breaking review integrity and influencing user behavior.

In addition, we consider attacks where legitimate users generate false reviews. As reviews are subjective in nature, it is difficult to determine whether the content of an authentic review is false or not. However, the TSE must prevent the sybil attacks which subvert the system by creating a large number of pseudonymous entities, using them to gain a disproportionately large influence. Since the TSE assigns multiple pseudonyms to a registered user, the sybil attacks can easily happen in the TSE as follows:

Sybil Attack 1: Such an attack is launched by malicious users: registered users leave multiple reviews toward a vendor in a time slot where reviews are false and negative to the service.

Sybil Attack 2: Such an attack is launched by malicious vendors with colluded users: a malicious vendor asks registered users to leave multiple reviews toward itself in a time slot where reviews are positive to the service.

The above two sybil attacks produce inaccurate information, which is unfair to either vendors or users, and disrupt the effectiveness of the TSE. In previous requirement, to prevent review linkability attacks, reviews are needed to be linked to real identities by the group authorities. However, the group authorities are not part of the network, and the detection of the sybil attacks by the group authorities is inefficient and probably with huge delay. To this end, we introduce another security mechanism to effectively resist the sybil attacks by restricting each user to generate only one review toward a vendor in a pre-defined time slot. If any user generates two or more than two reviews with different pseudonyms toward a vendor in a time slot, its real identity will be exposed to the public and such malicious behavior will be caught. The above two sybil attacks can thus be resisted.

Note that, restricting the number of reviews per each user in one time slot effectively limits the sybil attacks. However, any user can still generate false reviews using multiple pseudonyms for different time slots, and the reviews cannot be linked immediately. Since users are grouped based on their interests and reviews are linked to the social groups, false reviews will damage group reputation in a long run. Group reputation can therefore be taken as a weighting factor for the reviews generated by the group members. To further mitigate the effect by the false reviews, users may also make their service selection decisions based on reviews from familiar groups with high reputation rather than strange groups with low reputation.

4.3 TSE Solutions

We present the bTSE based on the above defined models. In the bTSE, a user, after being serviced by the vendor, submits a review to the vendor, which then records the review in its local repository. The review consists of two parts: (α, σ), where α is the review content and σ is the signature proving the authenticity of the content. Review submission may need cooperations from other users when the vendor is not in the transmission range of the user, or when direct submission fails due to communication failure. The logic is: the user forwards its review to a nearby user who wants to submit a review to the same vendor and expects that user to submit their reviews together. User cooperation increases submission rate and reduces submission delay at the cost of additional transmission efforts. To have a clear idea about the cost of user cooperation, we analyze the communication overhead with or without user cooperation being engaged.

Without cooperation, users submit only their own reviews, and the total communication overhead of l users (one review per user) is $l \cdot f(|\alpha| + |\sigma|)$, where $f(x)$ is the communication cost on transmitting x bits. With cooperations, in an extreme case, user u_{k_i} requires user $u_{k_{i+1}}$ to submit a review for it, $u_{k_{i+1}}$ further requires $u_{i_{i+2}}$, and so on, and $u_{k_{i+l-1}}$ finally submits the reviews of the l users altogether. The communication overhead is $(\sum_{j=1}^{l} j) \cdot f(|\alpha| + |\sigma|)$. If we further adopt the aggregate signature scheme [80], multiple signatures σ can be aggregated into a single batch signature σ^*, where σ^* has the same size as σ, and the communication overhead can be reduced to $(\sum_{j=1}^{l} j) \cdot f(|\alpha|) + l \cdot f(|\sigma|)$.

During review submission, data confidentiality and access control are not necessary because review information is open to the public, and data integrity, authenticity and non-repudiation can be obtained by directly applying traditional cryptography techniques such as hashing and digital signature on review content. As these techniques are very classic, we do not detail them here. While the basic security features are easy to achieve, it is challenging to resist the three review attacks and the two sybil attacks. To overcome this challenge, we first introduce the bTSE which uses tokens to synchronize review submission and organize reviews into certain structures. The integrity of the review structure is protected through hierarchical and aggregate signature techniques so that the review modification can be detected. User cooperation is further exploited to deal with the review rejection. Below, we elaborate on review structure, token usage, and the review generation and submission processes.

4.3.1 Phase 1 of bTSE: Structured Reviews

In the bTSE, reviews are structured to reflect their adjacency (i.e. submission order) through user cooperation. As such, vendors simply rejecting or modifying reviews will break the integrity of the review structure, thus being detected by the public. Consider a collection of n reviews received by a vendor v. We define four basic review structures (as illustrated in Fig. 4.3) and indicate vendors' review modification capabilities corresponding to them.

In Fig. 4.3a, reviews appear as discrete points, meaning that they are submitted separately and independent of each other. This independence gives the vendor maximum capability of manipulating the n reviews, and its modification capability is therefore $O(n)$. A logarithm modification capability is shown in Fig. 4.3b, where the reviews are presented in a tree-like structure. In this scenario, v is able to delete any single review corresponding to the leaf node, and the number of such reviews is $O(\log n)$. Figure 4.3c,d exhibits a chain structure and a ring structure.

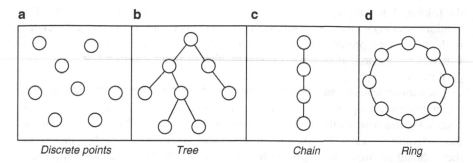

Fig. 4.3 Basic review structures

Fig. 4.4 A hybrid review
structure

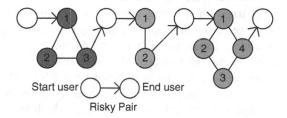

They respectively lead to constant $O(1)$ and zero modification capabilities. Clearly, the strength of the modification capabilities follows the order of $O(n) > O(\log n) > O(1) > 0$.

In order to restrict the vendor's review modification capability, reviews need to be structured. Pure use of the strongest ring structure requires extensive cooperation efforts from users, i.e., the first user that submitted a review must be aware of the pseudonyms of the users who are going to submit reviews subsequently. Considering the decentralized nature of the S-MSN, the assumption of having such pre-knowledge is unrealistic. Therefore, in the bTSE, separately submitted individual reviews and collectively submitted review clusters are linked into a chain according to their submission order, and within each review cluster, reviews are organized to form a ring. This hybrid structure, as shown in Fig. 4.4, limits the modification capability of the vendor below $O(1)$. Because this structure has a chain as its skeleton, in the sequel we refer to it as "chain" for ease of our presentation.

4.3.2 Phase 2 of bTSE: Token Generation

The chain structure requires reviews to be submitted sequentially. The bTSE uses a token technique to synchronize review submission. The vendor spontaneously initializes a number of tokens and issues them to distinct users, one per user. The tokens will then be circulated among users according to their local decision on token forwarding. A user cannot submit a review unless it currently holds one of the tokens. A token may be lost due to user mobility or malicious dropping. The vendor considers a token lost if it has not received any review submission associated to the token for a pre-defined maximum time duration θ_{exp}. It replaces lost tokens with new ones so as to maintain a constant number of active tokens and stable system performance.

Each token leads to an independent review chain. The vendor's review modification capability is proportional to the number of review chains. The more review chains, the less trustworthy the reviews from users' viewpoint. Thus, the vendor has the motivation to keep the token number as small as possible. On the other hand, there should be sufficient tokens in order to avoid token starvation problem, where some user never obtains a token to leave its review. In the performance evaluation, we will study the impact of token number on the system performance.

A user, when having a review to submit, transmits a token request message. After receiving the request, a nearby user currently holding a token or the vendor (if having a spare token) may send the token to the requesting user. The requesting user accepts the first arrived valid token and replies with an ACK message. For other received tokens, it replies with a RETURN message, indicating that it no longer needs a token. The token request, ACK and RETURN messages are signed by senders using (pseudonym) secret keys which are non-forgeable. Token forwarding happens toward one user at a time; successfully forwarded tokens (replied with ACKs) are no longer passed to any other user. Transmission retrials may be made up to a maximum number of times to tolerate communication failure.

The vendor maintains a token-pseudonym (TP) list. In this list, each token is linked to a pseudonym that belongs to a user who most recently submitted a review using the token. The list is updated whenever the vendor receives a new review, and is periodically broadcasted to all users in the vendor's transmission range. Once a token's information is published, the vendor cannot simply remove the token from the TP list because any modification to the list will cause inconsistency with previously published information and be noticed by the public. A user having a token will forward the token, after using it, to a randomly selected neighboring user who is wishing to submit a review. Below, we explain token structure and how a token is forwarded among users.

Consider three users u_1, u_2 and u_3, with u_1 neighboring u_2, and u_2 neighboring u_3. They are respectively members of groups g_1, g_2, g_3 and have obtained pseudonyms $pid_{1,1,*}, pid_{2,2,*}, pid_{3,3,*}$ from the corresponding group authorities. The vendor initializes a token with an identifier tok. It generates a public/private key pair (pk_t, sk_t) for tok and publishes the public key pk_t. Suppose that it intends to issue the token to u_1. Then, the token initially is a signature $\sigma_1 = Sign_{sk_t}(g_1||pid_{1,1,*}||T)$, where T is current time stamp. We denote this initial version σ_1 by tok_1. It implies that u_1 is the first user who can submit a review and must submit the review using pseudonym $pid_{1,1,*}$. The pseudonym $pid_{1,1,*}$ is exposed to the vendor by u_i.

After submitting a review using tok_1 and $pid_{1,1,*}$, u_1 updates tok_1 to tok_2 and passes tok_2 to u_2 as a response to u_2's token request. The updated version tok_2 is

$(PF_1, \sigma_2 = Sign_{psk_{1,1,*}}(g_2 || pid_{2,2,*} || T_1))$, where $PF_1 = (g_1, pid_{1,1,*}, \sigma_1)$ is the token forwarding proof of u_1. Note that, $(pk_t, tok, pid_{1,1,*})$ is currently included in the TP list. Suppose that tok_2 is the first token received by u_2. u_2 does the following: validate tok_2 by checking the authenticity of PF_1 using signatures σ_1 and σ_2, check if the user with $pid_{1,1,*}$ is the one that lastly forwards tok (by looking at the TP list), send an ACK to u_1, submit its review using tok_2 and $pid_{2,2,*}$, and update tok_2 to $tok_3 = (PF_1, PF_2, \sigma_3 = Sign_{psk_{2,2,*}}(g_3 || pid_{3,3,*} || T_2))$ where $PF_2 = (g_2, pid_{2,2,*}, \sigma_2)$, and send tok_3 to u_3.

The token forwarding process is repeated among users until tok expires or is brought out of the network. tok is always in the form of $(\{PF_x = (pid_{x,*,*}, \sigma_x)\}_{x \in \mathcal{X}}, \sigma_y)$ where u_x has forwarded the token and u_y the receiver user. It includes the hierarchical signatures that define the order of review submission and organizes submitted reviews in a chain structure. Note that malicious token drop is handled by the vendor through token replacement, as discussed previously.

Reducing Token Size by Signature Aggregation: We introduce an aggregate signature technique within multiple-authority settings, which is a variant of the scheme presented in [80]. This technique aggregates the signatures of different users from different social groups, and the signatures can be on different messages. By this technique, the signatures in a token can be aggregated, and the token size, thus the communication cost can be reduced. The aggregate signature technique will also be used for review aggregation in the next subsection, and the associated $Sign$ and $Verify$ functions will be instantiated as explained below.

Let $\mathbb{G}$ and $\mathbb{G}_T$ be two cyclic additive groups with the same prime order q, and $e : \mathbb{G} \times \mathbb{G} \to \mathbb{G}_T$ be a bilinear pairing (Sect. 1.4.4.1). P is a generator of $\mathbb{G}$. A group authority c_h picks a random $s_h \in \mathbb{Z}/q\mathbb{Z}$ and sets $Q_h = s_h P$. It also chooses two cryptographic hash functions $H_1 : \{0, 1\}^* \to \mathbb{G}$, and $H_2 : \{0, 1\}^* \to \mathbb{Z}/q\mathbb{Z}$.

Key Generation: A user u_j if registering to a group authority c_{h_j} will receive a bunch of pseudonym secret keys corresponding to randomly generated pseudonyms $pid_{j,h_j,*}$. Within a social group, the pseudonyms are never repeatedly assigned to users. The pseudonym secret keys $psk_{j,h_j,*} = (k_{j,0}, k_{j,1})$, where $k_{j,0} = s_{h_j} P_{j,0} = s_{h_j} H_1(pid_{j,h_j,*} || 0)$ and $k_{j,1} = s_{h_j} P_{j,1} = s_{h_j} H_1(pid_{j,h_j,*} || 1)$.

Signing: u_j generates a string as $str = $ "v", where v represents the identity of the vendor. Note that, all tokens are toward a specific vendor at a time period t. Therefore, the string can be obtained by other similar users. The signature on m_j will be $\sigma_j = Sign_{psk_{j,h_j,*}}(m_j) = (str, S_j, T_j)$.

$$S_j = r_j P_s + k_{j,0} + \beta_j k_{j,1} \text{ and } T_j = r_j P \tag{4.1}$$

where $P_s = H_1(str)$, $\beta_j = H_2(m_j, pid_{j,h_j,*}, str)$ and r_j is randomly chosen from $\mathbb{Z}/q\mathbb{Z}$.

Aggregation: Multiple signatures with the common str can be aggregated. Consider $\sigma_j = (str, S_j, T_j)$ for $1 \le j \le n$ are the signatures with common string str. The aggregated signature $\sigma_{agg} = (str, S_{agg}, T_{agg})$ can be obtained, where $S_{agg} = \Sigma_{j=1}^n S_j$ and $T_{agg} = \Sigma_{j=1}^n T_j$.

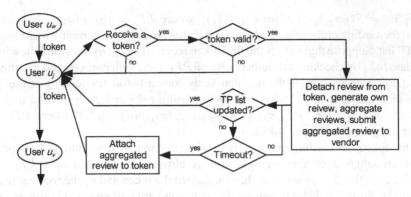

Fig. 4.5 Review generation and submission

Verification: Consider $\sigma_{agg} = (str, S_{agg}, T_{agg})$ is the aggregated signature for $\{(str, S_j, T_j)_{1 \le j \le n}\}$. The function $Verify(pid_{1,h_1,*}|| \cdots ||pid_{n,h_n,*}, m_1|| \cdots ||m_n, \sigma_{agg})$ outputs 1 if the following condition holds; 0 otherwise.

$$e(S_{agg}, P) \stackrel{?}{=} e(T_{agg}, P_s) \cdot$$
$$\Sigma_{j=1}^{n} e(H_1(pid_{j,h_j,*}||0) + \beta_j H_1(pid_{j,h_j,*}||1), Q_{h_j}) \tag{4.2}$$

where $\beta_j = H_2(m_j, pid_{j,h_j,*}, str)$. A user will only use $pid_{j,h_j,*}$ to generate a review on m_j for v only once to resist existential forgery attack [52].

4.3.3 Phase 3 of bTSE: Review Generation and Submission

Review generation and submission involve multiple steps as shown in Fig. 4.5. Review generation does not rely on tokens which gives users flexibility to generate review. Consider a user u_j who just received a token tok from a nearby user u_w with pseudonym $pid_{w,*,*}$. It checks if the received tok is valid. This validation step has two perspectives: (1) to ensure that tok is indeed originated from the vendor and has been properly forwarded in the past; (2) to ensure that tok is sent by the user who lastly used it. The first goal can be realized by using the public key pk_t of the vendor and the forwarder information (including secrets, pseudonyms, and time stamps) embedded in tok. The second one can be achieved by checking if the association $(tok, pid_{w,*,*})$ exists in the latest TP list provided by the vendor.

During token forwarding, a token is supposed to be passed to only one user that is wishing to submit a review to the same vendor. When multiple such users are present, a random selection can be made. In case that the token is passed to multiple users, whether accidentally (due to the failure in transmitting ACK message) or intentionally, the vendor will only accept the first subsequently submitted review using the token. With the second check on the TP list during token validation, the other users holding the token will find that the token is no longer valid and then try to find a new token to submit their reviews.

After confirming that tok is valid, u_j separates the attached review REV_w from tok. It checks the authenticity of REV_w. It is able to do so because u_w's pseudonym $pid_{w,*,*}$ is included in tok. If REV_w is invalid, u_j will discard it. After the review authenticity check, u_j generates its own review rev_j. Denote the review content by α_j. Suppose that u_j will use the pseudonym $pid_{j,h,*}$ from social group g_h for the review generation, and set T_j to current time which is larger than all the time stamps embedded in tok. It computes

$$\sigma_j = Sign_{psk_{j,h,*}}(\alpha_j||v||T_j)$$
$$rev_j = \langle g_h, pid_{j,h,*}, \alpha_j, v, T_j, \sigma_j \rangle. \tag{4.3}$$

The signature σ_j can be verified by checking $Verify(pid_{j,h,*}, \alpha_j||v||T_j, \sigma_j) \stackrel{?}{=} 1$ (see the previous subsection for the details of functions $Sign$ and $Verify$). The receiver then knows that rev_j is indeed generated by a user from g_h at time T_j, not forged by the vendor or a user from a different group. Having generated rev_j, u_j aggregates it with REV_w (by the signature aggregation technique in Sect. 4.3.2) and submits the aggregated reviews REV_j ($REV_j = rev_j$ if $REV_w = null$) together with tok to the vendor. The vendor checks the validity of REV_j and tok, and broadcast the updated TP list. Review aggregation is the same process as signature aggregation during token forwarding presented in the previous section. Review aggregation has two advantages: (1) it effectively resists the review attacks; (2) it largely reduces the communication overhead.

Note that u_j is unable to forge a review of u_w because it cannot obtain any pseudonym secret key $psk_{w,*,*}$, and u_j is unable to replace the review with any other review received from u_w in the past because time stamp is used to prevent review replay. Direct replacement can be easily detected and rejected by the vendor. Further, u_j cannot forward the token without submitting REV_w and/or rev_j because the token records the forwarding history and the vendor will detect the review missing when it later receives the token as part of a review submission made by another user.

After submitting REV_j and tok to the vendor, u_j checks the updated TP list from the vendor. An unsuccessful submission can be due to communication failure or review rejection. To tolerate communication failure, a number of submission retrials can be made before drawing a submission failure conclusion. Upon receiving the updated TP list, u_j will check which pseudonym tok is related to in the list. If tok is related to $pid_{j,h,*}$, meaning that u_j have successfully submitted REV_j, u_j will forward tok to a nearby user as described in the previous section. If tok is still related to $pid_{w,*,*}$, meaning that u_j's submission failed, u_j will resort for cooperative submission by sending tok and REV_j together to a nearby user that is requesting for a token. If tok is related to a different pseudonym, implying that u_w must have sent the token to multiple users and u_j's submission failed, u_j will try to find a new token from nearby users and submit REV_j using it.

Comments: During service information dissemination, the vendor needs to broadcast its entire review collection together with the latest versions of the tokens. After receiving the service information, a user checks the authenticity of the reviews and compares the pseudonyms associated with reviews to those embedded in tokens. Because the token contains its circulation history (implemented by hierarchical signatures and time stamps), the user may arrange the reviews according to the circulation history. Any missed review will be detected. If multiple reviews from the same user appear, it will use time stamp to differentiate them.

4.3.4 SrTSE

In previous subsections, we have introduced a bTSE where review linkability, rejection and modification attacks are considered. We further extend the bTSE to a Sybil-resisted TSE, named SrTSE, which effectively prevents the sybil attacks. In the following, we first describe the sybil attacks in the S-MSN, and then introduce our solutions to complete the design of the SrTSE.

Sybil Attacks: In Sect. 4.2.2, we define two types of sybil attacks: the sybil attack 1 is launched by a group of registered users. They aim at telling other users the bad service from a vendor while the service of the vendor is good. With the valid registration, these malicious users are able to leave false reviews toward a specific vendor. Even realizing the reviews are not in accord with the service, the vendor cannot simply delete or reject the reviews. If the vendor does, users will detect such behavior and regard the vendor as a dishonest vendor. Besides, the sybil attack 2 is launched by a vendor and a group of registered users. They aim at raising the reputation of the service from a vendor while the service of the vendor is not that good. The reviews generated by these malicious users cannot be distinguished from other reviews by well-behaving users. In the bTSE, every user receives multiple pseudonyms and the corresponding secret keys. For example, u_j has $pid_{j,h,1}, pid_{j,h,2}, \cdots$ in social group g_h. Since these pseudonyms are random numbers and cannot be linked by anyone except group authorities, u_j can use $pid_{j,h,1}, pid_{j,h,2}, \cdots$ to generate multiple reviews toward a vendor for a short time period. In addition, u_j can form the false reviews in chain structure or ring structure. Therefore, from the perspective of other users, they cannot tell if these reviews are from the same user or not.

Sybil-Resisted TSE (SrTSE): In the SrTSE, we introduce a novel solution to prevent the two sybil attacks. In the S-MSN, we consider that a user has no need to generate multiple reviews toward a vendor in a short time period. The SrTSE allows a user to leave only one review toward a vendor for a pre-defined time slot. If a user generates multiple reviews with the same pseudonyms, the linkability of the reviews can be easily verified by the public; if a user generates multiple reviews with different pseudonyms toward a vendor in a time slot, its real identity will be exposed to the public. To achieve the above properties, we modify the pseudonym generation and the signature scheme of the bTSE.

Let $\mathbb{G}$ and $\mathbb{G}_T$ be two cyclic additive groups with the same order q, and e : $\mathbb{G} \times \mathbb{G} \to \mathbb{G}_T$ be a bilinear pairing [78]. P, Q are two generators of $\mathbb{G}$. A group authority c_h picks a random $s_h \in \mathbb{Z}/q\mathbb{Z}$ and sets $Q_h = s_h P$. It also chooses two cryptographic hash functions $H_1 : \{0, 1\}^* \to \mathbb{G}$, and $H_2 : \{0, 1\}^* \to \mathbb{Z}/q\mathbb{Z}$.

Consider a user u_j registers to the social group g_h in the SrTSE. Then, u_j obtains the following values:

- $pid_{j,h_j,*}$, a published random number.
- $a_{j,*} = pH_2(pid_{j,h_j,*}) + id_j$, where id_j is the real identity of u_j, and ρ is a coefficient in $\mathbb{Z}/q\mathbb{Z}$.
- $b_{j,*} = (r_* P, s_h Q + r_* H_1(a_{j,*} r_* P \| pid_{j,h_j,*}))$, where r_* is a random number. This is a signature on $a_{j,*} r_* P$ by the group authority c_h.

For multiple random numbers $pid_{j,h_j,*}$, u_j obtains multiple tuples ($pid_{j,h_j,*}, a_{j,*}, b_{j,*}$) from c_h. Then, u_j regards $pid_{j,h_j,*}$ as the pseudonym and $psk_{j,h_j,*} = a_{j,*}$ as the secret key. In order to generate a signature on message m_j, u_j executes the following steps:

- u_j calculates $a_{j,*} H_1(m_j)$.
- u_j generates a random number $\bar{r} \in \mathbb{Z}/q\mathbb{Z}$, and outputs a signature

$$\sigma_j = (pid_{j,h_j,*}, \sigma_{j,1}, \sigma_{j,2}, \sigma_{j,3}, \sigma_{j,4}), \tag{4.4}$$

where $\sigma_{j,1} = a_{j,*} H_1(m_j)$,

$$\begin{aligned} \sigma_{j,2} &= a_{j,*} r_* P, \\ \sigma_{j,3} &= r_* P, \\ \sigma_{j,4} &= s_h Q + r_* H_1(a_{j,*} r_* P \| r_* P \| pid_{j,h_j,*}). \end{aligned} \tag{4.5}$$

If an entity receives σ_j, it runs the following verification steps.

- Step 1:

$$e(\sigma_{j,4}, P) = e(Q, Q_h) e(H_1(\sigma_{j,2} \| \sigma_{j,3} \| pid_{j,h_j,*}), \sigma_{j,3}) \tag{4.6}$$

- Step 2:

$$e(\sigma_{j,1}, \sigma_{j,3}) = e(H_1(m_j), \sigma_{j,2}) \tag{4.7}$$

It can be seen that $(\sigma_{j,3}, \sigma_{j,4})$ is a signature generated by the group authority c_h since s_h is the secret key only known to c_h. From Step 1, the authenticity of $\sigma_{j,2}$ and $pid_{j,h_j,*}$ can be guaranteed. In addition, from Step 2, if the equality holds, it is publicly verified that u_j knows the value $a_{j,*}$. In fact, we build our signature scheme based on identity-based signature [116] and short signature [117].

Sybil Attack Detection: For each review, we require users to sign on $m_j = v\|t$ where v is the vendor's name and t is the time slot. If users do not output the signature on m_j, its review will not be accepted by the public. We consider a sybil attack launched by u_j who generate two reviews with two different pseudonyms $pid_{j,h_j,1}$ and $pid_{j,h_j,2}$. If both reviews are authentic, they must contain both $a_{j,1}H_1(m_j)$ and $a_{j,2}H_1(m_j)$ which can be accessed by the public. Thus, the public is able to calculate $Tr = id_j H_1(m_j)$ from

$$a_{j,1} = \rho H_2(pid_{j,h_j,1}) + id_j \tag{4.8}$$

and

$$a_{j,2} = \rho H_2(pid_{j,h_j,2}) + id_j, \tag{4.9}$$

since

$$id_j = \frac{a_{j,1}H_2(pid_{j,h_j,2}) - a_{j,2}H_2(pid_{j,h_j,1})}{H_2(pid_{j,h_j,2}) - H_2(pid_{j,h_j,1})}. \tag{4.10}$$

To recover the real identity of the sybil attacker, any entity calculates $Tr' = idH_1(m_j)$ for every possible id and tests if $Tr' \overset{?}{=} Tr$. The entity outputs the recovered identity id, upon satisfaction of the above equation.

Note that, similar to [53, 54], the vendors or the users can pre-calculate values $idH_1(m_j)$ for every possible identity, and then, they just need to check the equality between Tr and these values. Within a constant time, the real identity of the sybil attacker can be revealed.

Aggregate Signature in the SrTSE: The signature aggregation plays an important role in the bTSE because it largely reduces the communication overhead. We will also explore the possible aggregation scheme for the newly developed signatures in the SrTSE. Observing the modified signature scheme, the pseudonyms and the corresponding secret keys have to be equipped with a trapdoor such that other entity (not group authority) is able to recover the real identity of the sybil attacker. Therefore, the aggregation on signatures becomes more difficult. From the verification Step 1 and Step 2, we can see that $\sigma_{j,1}, \sigma_{j,2}$ and $\sigma_{j,3}$ cannot be aggregated because $\sigma_{j,2}$ and $\sigma_{j,3}$ have to be individually input in the hash function and $\sigma_{j,1}$ is paired with different $\sigma_{j,3}$ every time. But $\sigma_{j,4}$ from different users can be aggregated in the form of $\prod_j \sigma_{j,4}$ because it is always paired with P. The verification on the aggregate signature is shown below.

$$e(\prod_j \sigma_{j,4}, P) = e(Q, Q_h) \prod_j e(H_1(\sigma_{j,2}\|pid_{j,h_j,*}), \sigma_{j,3}) \tag{4.11}$$

In the SrTSE, we have modified the signatures of the review generation in order to resist the sybil attacks. Such modification makes the aggregation less efficient.

Table 4.1 Comparison of security

	L_att	R_att	M_att	S_att
Non-Coop	Y	N	N	N
bTSE [5]	Y	Y	Y	N
SrTSE	Y	Y	Y	Y

Table 4.2 Comparison of communication overhead

	S_token_1	S_token_k	S_review_1	S_review_k
bTSE [5]	$2\|\mathbb{G}\|$	$2\|\mathbb{G}\|$	$2\|\mathbb{G}\|$	$2\|\mathbb{G}\|$
SrTSE	$2\|\mathbb{G}\|$	$2\|\mathbb{G}\|$	$4\|\mathbb{G}\|$	$(3k+1)\|\mathbb{G}\|$

If we have n signatures and each has four group elements in $\mathbb{G}$, these signatures can be aggregated into $3k + 1$ group elements. Though the size of aggregate signature decreases, it still linearly depends on k. We regard the size of such aggregate signature as $O(k)$. In comparison, the bTSE offers an aggregated signature sized at $O(1)$. Thus, we can still use the aggregate signature scheme of the bTSE for the token generation to achieve higher efficiency.

4.3.5 Summary of bTSE and SrTSE

We have introduced two trustworthy service evaluation systems: one considers the review attacks only and the other one considers both the review attacks and the sybil attacks. In the following, we summarize the efficiency and security properties of these two systems. We also consider the non-cooperative system where pseudonyms are employed and the reviews are individually submitted by users. Let "L_att" denote "review linkability attacks", "R_att" denote "review rejection attacks", "M_att" denote "review modification attacks", "S_att" denote "sybil attacks", "S_token_1" denote "the size of signature on tokens", "S_review_1" denote "the size of signature on reviews", "S_token_k" denote "the size of k-aggregated signatures on tokens", "S_review_k" denote "the size of k-aggregated signatures on reviews", "Y" denote "resist", and "N" denote "not resist".

From the above the security comparisons in Table 4.1, it can be seen that both the bTSE and the SrTSE outperforms the non-cooperative system in terms of security. Moreover, the bTSE resists "L_att", "R_att", and "M_att" which are the possible attacks. The SrTSE additionally resists the two sybil attacks from malicious users, and thus the SrTSE is more reliable and trustworthy in the S-MSN.

In the above Table 4.2, we give the communication overhead of a token and a review in both bTSE and SrTSE. We do not count the sizes of messages and the common strings because their sizes are negligible compared to the signatures. From Table 4.2, both bTSE and SrTSE have very efficient review and token generation due to the signature aggregation. To resist the sybil attacks, SrTSE employs a trapdoor in the pseudonym which leads to a linearly-increasing size in review generation of the SrTSE.

4.4 Security Analysis

Security analysis focuses on the system's resilience against review linkability attacks, review rejection attacks, review modification attacks, and sybil attacks.

4.4.1 Resilience to Review Linkability Attacks

In a review linkability attack, a user submits unlinkable reviews. If reviews without linkability enabled on the group authorities are allowed, malicious users may abuse their memberships and generate irresponsible reviews to undermine the system's performance. Recall the review generation process described in Sect. 4.3.3. A review rev_j is valid if and only if the following verification function can be checked by the public: $Verify(pid_{j,h,*}, \alpha_j||v||T_j, \sigma_j)$. By the verification function, $(g_h, pid_{j,h,*}, \alpha_j||v||T_j)$ are related by a non-forgeable signature. Anyone without the secret key sk_h or $psk_{j,h,*}$ is unable to forge a triple tuple in such relation. Furthermore, when generating $pid_{j,h,*}$ and the corresponding $psk_{j,h,*}$ for user u_j, the group authority c_h checks the unique identity id_j of u_j, and maintains an association $(pid_{j,h,*}, id_j)$ all the time. Therefore, invalid reviews are recognizable by the public and the group authorities, and the group authorities are able to link any valid review to the unique identity of its generator. Review linkability attacks thus can be effectively prevented.

4.4.2 Resilience to Review Rejection Attacks

In a review rejection attack, the vendor rejects unwelcome but authentic reviews from users. Recall the review submission process in Sect. 4.3.3. A user u_j tries to directly submit its review rev_j using token tok to the vendor for several times. If all the trials fail, whether due to communication failure or owing to a review rejection attack, it will pass both tok and rev_j to a nearby user u_k who also has a need in submitting its review rev_k to the same vendor, and expect u_k to submit rev_j and rev_k together. Then, u_k validates the received tok and rev_j, aggregates rev_j and rev_k to obtain REV_k, and submits REV_k as a whole, together with tok, to the vendor. The vendor either accepts REV_k (including the previously rejected rev_j) or rejects it (including the new one rev_k). Now, the vendor has a constraint on rejecting REV_k because it has to consider whether rev_k is complimentary or not. As reviews are aggregated, the vendor will have more and more constraints on launching review rejection attacks. If it finally decides to accept the aggregated reviews that are submitted as a whole piece, it will actually accept all the reviews that it previously rejected. The review aggregation and indirect review submission techniques mitigate such attacks.

4.4.3   Resilience to Review Modification Attacks

In a review modification attack, the vendor manipulates its locally recorded review collection by inserting forged complimentary reviews and modifying or deleting existing unwelcome authentic reviews. The integrity of the review content is guaranteed by the signature techniques (refer to Sect. 4.4.1). Reviews are generally linked in one or a few review chains, depending on the number of the used tokens, if they are directly submitted by users. The cooperation among users enables indirect review submission in case of direct-submission failure. Indirect submission causes the reviews from different users to be aggregated and formed in a ring structure. Figure 4.4 shows a single review chain as a result of direct and indirect submissions. In this figure, dots represent individual reviews, and without ambiguity they also refer to the users that submit the corresponding reviews.

Let the users whose reviews are aggregated in one signature form a cluster. In Fig. 4.4, there are three clusters of users as indicated by the colorful dots. We index the users in each cluster such that a smaller index indicates the user obtains the token earlier and the largest index implies the user successfully and directly submitted the aggregated reviews to the vendor. Thus, the users with the smallest index and the largest index are the interfaces of the cluster. Outside these clusters, arrowed lines indicate the token forwarding direction. We define a risky pair of users as two users that are interconnected by an arrowed line. The one from which the line starts is called start user, and the other is referred to as end user. The following theorems indicate that review modification attack can be resisted.

Theorem 4.1. *The vendor is able to insert reviews into a review chain if and only if it has compromised a start user.*

Proof. Suppose that the vendor has compromised a start user u_j and obtained its all keying materials. Let u_v be the end user corresponding to u_j. The token is in the form of $(\cdots, PF_j, PF_v, \cdots, \sigma_y)$. If the vendor wants to insert a false review via the pseudonym of user u_m, it has to change the token to $(\cdots, PF'_j, PF_m, PF_v, \cdots, \sigma_y)$, where

$$PF'_j = (pid_{j,h_j,*}, Sign_{psk_{j,h_j,*}}(g_{h_m} || pid_{m,h_m,*} || T_j)),$$
$$PF_m = (pid_{m,h_m,*}, Sign_{psk_{m,h_m,*}}(g_{h_v} || pid_{v,h_v,*} || T_m)). \tag{4.12}$$

The validity of the modified token can be easily verified. Note that in this case, the vendor has also compromised u_m because otherwise it would not have $psk_{m,h_m,*}$ and not be able to generate PF_m or forge the review.

We consider the case that the vendor has successfully inserted a forged review. In order for the forged review to be accepted by the public, the vendor must have inserted a fake token forwarding proof in the token, which in turn implies that it must have compromised a user who has used the token. Assume that the compromised user is not a start user. In this case, it must be a user in a cluster (see Fig. 4.4) that does not have the largest index. Because the user with largest index outputs a non-forgeable aggregated signature on the legitimate reviews in the cluster, the forged review will be detected, contradicting to the fact that the insertion is successful. This completes the proof.   □

Theorem 4.2. *The vendor without compromising any user can only delete a sub-chain of a review chain, starting with an end user and spanning all the users that receive the token later than the end user.*

Proof. A cluster of aggregated reviews are treated as a single piece because the vendor is not able to separate them. They are either all kept or all deleted. By definition, an end user is also a start user unless it is within a user cluster (corresponding to a cluster of aggregated reviews). Thus, the vendor can only delete reviews from an end user. After deleting a review or some aggregated reviews, the review chain becomes broken. The breakage is detectable unless the subsequent reviews are also deleted. However, if the vendor compromises a start user (and obtains the user's keying materials), it will be able to delete an arbitrary number of successive reviews including review clusters from the start user, and fix the chain breakage, without being detected.   □

4.4.4 Resilience to Sybil Attacks

To resist to the sybil attacks, we need to prove that the SrTSE satisfies the following two properties.

- *P1*. If a user leaves two or more false reviews with different pseudonyms toward a vendor in a time slot, its real identity can be derived by the vendor and other users.
- *P2*. If a user leaves only one review toward a vendor in a time slot, its real identity can be protected.

We first consider the property *P1* of the SrTSE. We consider a malicious user u_j generates two signatures on $m_j = v\|t$. From the signature, $\sigma_{j,1}$ can be obtained. If both signatures are valid, the relation $\sigma_{j,1}$, $\sigma_{j,2}$ and $\sigma_{j,3}$ will be fixed by Eq. (4.7). Since $\sigma_{j,2}$ and $\sigma_{j,3}$ are included in the message of $\sigma_{j,4}$, their authenticity can be verified. Therefore, $\sigma_{j,1} = a_{j,*}H_1(m_j)$ where $a_{j,1}$ is generated in the specific format by the group authority. The public can obtain $a_{j,1}H_1(m_j)$ and $a_{j,2}H_1(m_j)$, and then derive

$$id_j H_1(m_j) = \frac{H_2(pid_{j,h_j,2})a_{j,1} - H_2(pid_{j,h_j,1})a_{j,2}}{H_2(pid_{j,h_j,2}) - H_2(pid_{j,h_j,1})} \cdot H_1(m_j) \qquad (4.13)$$

By executing the equality checks, the real identity id_j of u_j will be determined. Note that, ρ is determined by the group authorities. Different group generate different ρ. We consider the used two pseudonyms $pid_{j,h_j,1}$ and $pid_{j,h_j,2}$ are from the same social group in the above analysis. We can further require a trusted third authority to coordinate all the group authorities to generate the same ρ for one user, and then the sybil attacks using pseudonyms from different groups can be resisted.

We then consider the property $P2$ of the SrTSE. From a signature σ_j, the real identity can be disclosed from $a_{j,*}$ which is contained in $\sigma_{j,1} = a_{j,*}H_1(m_j)$ and $\sigma_{j,2} = a_{j,*}r_*P$. Denote $H_1(m_j) = r'P$. Thus, we have two values

$$\sigma_{j,1} = (\rho H_2(pid_{j,h_j,*}) + id)r'P \tag{4.14}$$

and

$$\sigma_{j,2} = (\rho H_2(pid_{j,h_j,*}) + id)r_*P. \tag{4.15}$$

If multiple signatures with different pseudonyms are generated toward different m_j by u_j, we can obtain the following values:

$$\begin{aligned}
(\rho H_2(pid_{j,h_j,1}) + id_j)r'_1P, \\
(\rho H_2(pid_{j,h_j,1}) + id_j)r_1P, \\
(\rho H_2(pid_{j,h_j,2}) + id_j)r'_2P, \\
(\rho H_2(pid_{j,h_j,2}) + id_j)r_2P, \cdots
\end{aligned} \tag{4.16}$$

Since $(r'_1, r_1, r'_2, r_2, \cdots)$ are independent and unknown to the public. The random number ρ cannot be removed by the linear combination of these values. Therefore, the real identity id_j is always anonymized by ρ, and thus id_j is protected.

4.4.5 Numerical Results of Detecting Sybil Attack

The SrTSE can resist the sybil attack, i.e., the sybil attack can be detected without the involvement of the group authorities. In the following, we study the performance of the SrTSE under the sybil attack. We will evaluate how much computation costs needed to detect the false reviews by the sybil attack. We first consider the case of a single malicious user in the SrTSE. The sybil attacker generates x false reviews toward the vendor in time slot t using its x different pseudonyms. The vendor totally receives y reviews in time slot t ($y \geq x$). From Eq. (4.13), the vendor needs to do every calculation for any pair of the received reviews. That means, the maximum number of calculation is $\binom{y}{2}$. We denote the number of calculations needed to filter all the false reviews by $C_1(x, y)$ ($\leq \binom{y}{2}$). In fact, if two reviews have been identified to be associated with the attacker, all the rest false reviews can be easily identified by comparing them with the detected false reviews. Thus, the expected value of $C_1(x, y)$ is calculated in Eq. (4.17).

$$C_1(x, y) = \frac{y - x}{y}(y - 1 + C_1(x, y - 1)) + \frac{x}{y}(y - 1)$$

$$= y - 1 + \frac{y - x}{y}C_1(x, y - 1) \tag{4.17}$$

In the above equation, $\frac{y-x}{y}$ and $\frac{x}{y}$ represent the probabilities of choosing a valid review and a false review, respectively. If a valid review is chosen, we need to do $y - 1$ calculations between the chosen review with the rest $y - 1$ reviews and $C_1(x, y - 1)$ calculations among $y - 1$ reviews. If a false review is chosen, we need to do the first $y - 1$ calculations and then all the false reviews will be detected. Similarly, we further derive the number of calculations $C_2(x_1, x_2, y)$ in case of two malicious users, as shown in Eq. (4.18), where x_1 and x_2 represent the numbers of false reviews of two malicious users, respectively. We have $x_1 + x_2 \leq y$.

For the initial values, we have $C_1(x, x) = x - 1$, $C_2(0, x_2, y - x_1) = C_1(x_2, y - x_1)$ and $C_2(x_1, 0, y - x_2) = C_1(x_1, y - x_2)$. If $x_1 + x_2 = y$, $C_2(x_1, x_2, y) = y - 2 + \frac{2x_1x_2}{y}$.

$$C_2(x_1, x_2, y) = \frac{y - x_1 - x_2}{y}(y - 1 + C_1(x_1, x_2, y - 1))$$

$$+ \frac{x_1}{y}(y - 1 + C_2(0, x_2, y - x_1))$$

$$+ \frac{x_2}{y}(y - 1 + C_2(x_1, 0, y - x_2)) \tag{4.18}$$

$$= y - 1 + \frac{y - x_1 - x_2}{y}C_1(x_1, x_2, y - 1)$$

$$+ \frac{x_1}{y}C_2(0, x_2, y - x_1) + \frac{x_2}{y}C_2(x_1, 0, y - x_2)$$

Then, we plot $C_1(x, y)$ and $C_2(x_1, x_2, y)$ and $C_2(x, y)$ in terms of x, y, x_1 and y_1, respectively, as shown in Fig. 4.6. From Fig. 4.6a, in case of 1 malicious user, it can be seen that the number of calculations almost increases linearly as the number of received reviews increases. When more reviews received at the vendor, more calculation efforts are needed to find the false reviews. Moreover, when the number of false reviews increases, the calculation efforts can be reduced because the probability of finding a false review is larger. From Fig. 4.6b,c, we can observe that when the number of false reviews decreases or the number of received reviews increases, the number of calculations to detect all false reviews increases. Note that when $x_1 = x_2 = 15$ and $y = 30$, the number of calculations is 43. In this case, 30 reviews are all false reviews, and we still need 43 calculations on average to detect them. The reason is that the calculations cannot detect any false reviews when the two reviews are separately from two users.

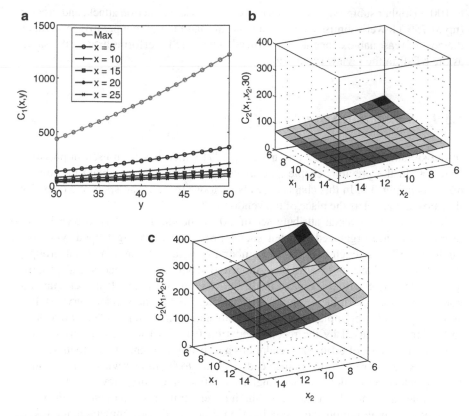

Fig. 4.6 Number of calculation on detecting the sybil attack. (**a**) 1 malicious user. (**b**) 2 malicious users, 30 reviews. (**c**) 2 malicious users, 50 reviews

4.5 Performance Evaluation

In this section, we evaluate the performance of the bTSE through trace-based custom simulations. We choose to compare the bTSE with a non-cooperative system, where each user directly submits its review to the vendor without any synchronization constraint (use of tokens). We use the following three performance metrics:

- *Submission rate*: It is defined as the ratio of the number of successfully submitted reviews to the total number of generated reviews in the network.
- *Submission delay*: It is defined as the average duration between the time when a review is generated and the time when it is successfully received by the vendor.
- *Cooperation overhead*: It is defined as the total number of times that tokens are forwarded among users.

Because the non-cooperative system involves only direct review submission, the last metric is not applicable to it. As we will see, the bTSE achieves significantly (up to 100%) higher submission rate under a defined review rejection attack, and greatly (up to 75%) lower submission delay in general than the non-cooperative system, at the cost of reasonable cooperation overhead. The SrTSE performs exactly the same as the bTSE in the review submission.

4.5.1 Simulation Setup

In the simulation, we use the real trace log [77] which has been used in previous chapters. In the previous chapter, we have shown that how to obtain the top 10 hotspots in Fig. 3.8. In this chapter, we also choose these 10 hotspots and consider the chosen hotspot as the place of the vendor.

We define a universal attribute set of 50 elements. The set is known by all users. Users are organized into ten social groups, each being tagged with five random attributes. Each user has a membership with 1–5 random social groups, that is, it may have 5–25 attributes, inherited from the belonged social groups. The vendor (precisely, its service) has three random attributes. If a user shares a common attribute with the vendor, it will be interested in the vendor (service). For simplicity, we do not implement users random state transition from "not interested" to "interested" caused by the recommendation from its friends. Each user has a transmission range of 80 m. The vendor has a transmission range equal to its service range (SR). A user interested in the vendor wishes to submit a review to the vendor when it enters the vendor's service range for the first time. Direct review submission is possible only when the vendor is within the user's transmission range. As the trace log covers a small region and a small period time, we do not implement the token timeout interval θ_{exp} (see Sect. 4.3.2).

We conduct two sets of simulations under the situations with/without review attacks. We vary SR between 150 and 300 m, and token number TN between 1 and 10. As analyzed in Sect. 4.4, the bTSE resists the review linkability and modification attacks through cryptography techniques and specially-designed review structure, and mitigates review rejection attack through cooperative review submission. The first two attacks have no influence on review submission. In our simulation study, we are therefore interested only in the impact of review rejection attack on the system performance. Each review is a value ranged in [0, 1]. A review is negative if its value is lower than 0.5. The vendor performs review rejection action by rejecting all negative reviews. When multiple reviews are aggregated and submitted together, the vendor accepts them all if their average value is no less than 0.5, or rejects them all otherwise. We place the vendor at the centers of the 10 hotspots in turn and conduct 50 simulation runs for each placement. Using the total 500 simulation runs, we obtain the average results to be analyzed in the next subsection.

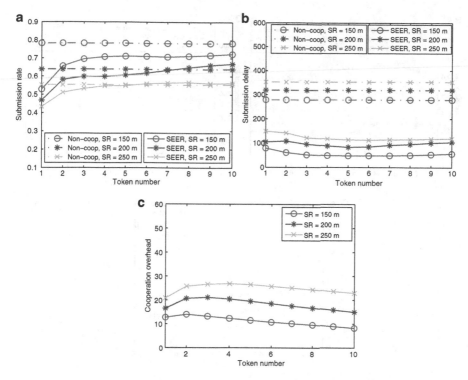

Fig. 4.7 Performance under no review rejection attack. (**a**) Submission rate. (**b**) Submission delay. (**c**) Cooperation overhead

4.5.2 Simulation Results

Under no Review Rejection Attack: We first study the system performance in relation with SR. Let us observe Fig. 4.7. When SR goes up, the number of users who enter the service range and thus generate reviews increases. Recall that each user has a transmission range much smaller than SR. In the non-cooperative system, users have to move close enough to the vendor in order to submit their reviews. Hence, the system shows a decreasing submission rate and increasing submission delay with SR. In the bTSE, review submission is constrained by token possession in addition to user-to-vendor distance on one hand. On the other hand, cooperative review submission is triggered when direct submission is not possible. The interplay of the two factors renders the bTSE exhibiting a performance trend similar to the non-cooperative system's in submission rate and submission delay as SR varies. From Fig. 4.7b, the bTSE has greatly lower submission delay than the non-cooperative

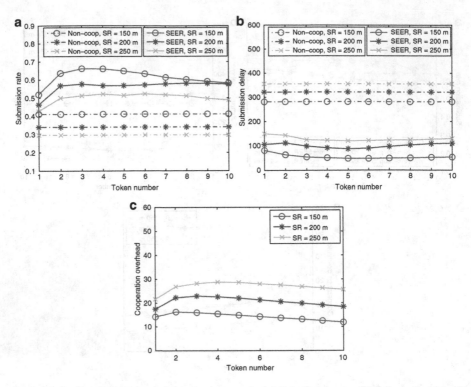

Fig. 4.8 Performance under review rejection attack. (**a**) Submission rate. (**b**) Submission delay. (**c**) Cooperation overhead

system, up to 75% lower. Figure 4.7c depicts the cooperation overhead of the bTSE. As expected, the larger the vendor's service range, the more cooperation efforts from users involved.

We then look at how TN impacts the system performance. Intuitively, when TN goes up, users have increased opportunity to submit reviews, leading to raised system performance. This intuition is confirmed by the results in Fig. 4.7a,b. We observe an arguable phenomenon: submission rate and delay both stabilize after TN is beyond certain value. In the case of SR = 150, it occurs after TN = 20 and is however not shown here. The reason for this phenomenon is as follows. When there are more tokens circulating in the network, initially users can easily get tokens and submit their reviews. Recall that users no longer participate in the review system once their reviews are submitted to the vendor or forwarded to others. Over time, the network of participating users becomes sparse and sparse, and these users have less and less chance to receive a token due to decreased network density. This can be cross verified by the cooperation

Under Review Rejection Attack: Figure 4.8a,b shows the performance comparison of the bTSE and the non-cooperative system when the vendor launches the review rejection attack. We observe that the non-cooperative system has a performance drop (>25%) in submission rate. Indeed, it is not equipped with any security mechanism against the attack and suffers performance degradation. Submission delay does not show any noticeable change since only direct submission is engaged in the non-cooperative system and only successfully submitted reviews are considered during delay calculation. Compared with the case of no review rejection attack, the bTSE only has slightly reduced (<10% smaller) submission rate and nearly unchanged submission delay thanks to the user cooperation and review aggregation mechanisms. The bTSE achieves significantly higher submission rate than the non-cooperative system, up to 100%. These simulation results indicate that the bTSE can effectively resist the review rejection attack.

4.6 Related Work

Trust evaluation of service providers is a key component to the success of location based services in a distributed and autonomous network. Location-based services require a unique and efficient way to impress the local users and earn their trust so that the service providers can obtain profits [118, 119]. Rajan and Hosamani [120] used an extra monitor deployed at the untrusted vendor's site to guarantee the integrity of the evaluation results. Wang and Li [121] proposed a two-dimensional trust rating aggregation approach to enable a small set of trust vectors to represent a large set of trust ratings. Ayday and Fekri [122] approached the trust management as an inference problem and proposed a belief propagation algorithm to efficiently compute the marginal probability distribution functions representing reputation values. Das and Islam [123] introduced a dynamic trust computation model to cope with the strategically altering behavior of malicious agents.

Distributed systems are vulnerable to *sybil attacks* where an adversary manipulates bogus identities or abuse pseudonyms to compromise the effectiveness of the systems. For example, in the peer-to-peer networks, Douceur [124] indicated that the sybil attacks can compromise the redundancy of distributed storage systems. In the sensor networks, Karlof and Wagner [125] showed that the sybil attacks can damage the routing efficiency. Newsome et al. [126] proposed many defense mechanisms, such as, radio resource testing, key validation for random key pre-distribution, and position verification. In vehicular ad hoc networks, Lu et al. [127] proposed an efficient detection mechanism on double-registration which can be conducted to mitigate the possible sybil attacks. The sybil attacks in social networks have attracted great attention recently [56–58]. In social networks, Wei et al. [58] mentioned the existence of a trusted authority can mitigate the effect of the sybil attacks, but they considered that such requirements impose additional burdens on users which is not acceptable.

4.7 Conclusion and Future Directions

We have introduced a trustworthy service evaluation (TSE) system for service-oriented mobile social network (S-MSN). The TSE enables a vendor to receive, record the reviews from its customers and disseminate the reviews to other nearby users. It helps build the user-to-vendor trust from the user-to-user trust. Specifically, the TSE engages hierarchical signature and aggregate signature techniques to transform independent reviews into structured review chains. This transformation involves distributed user cooperation, which improves review integrity and greatly reduces vendors' modification capability. We have presented three review attacks and shown that the bTSE can effectively resist the review attacks without relying on a trusted authority. We have also considered the notorious sybil attacks and demonstrated that such attacks cause huge damage to the bTSE. We have subsequently modified the construction of pseudonyms and the corresponding secret keys in the bTSE, and obtained a sybil-resist TSE (SrTSE) system. The SrTSE allows users to leave only one review toward a vendor in a pre-defined time slot. If multiple reviews with different pseudonyms from one user are generated, the real identity will be disclosed to the public. The sybil attack is thus prevented in the SrTSE. Numerical results show the effectiveness of the SrTSE to resist the sybil attacks. Further trace-based simulation results demonstrate that the bTSE significantly outperforms a non-cooperative review system in terms of submission rate and delay, especially in the presence of the review rejection attacks.

The TSE systems in the MSN still have many challenging design issues. In the distributed MSN, users and vendors are possible attackers. There is no trusted entities that can be reached in real-time. Vendors need to regenerate tokens to prevent the attacks from users, while users need to do the cooperation to prevent the review attacks from vendors. We build the TSE systems without directly relying on the credit or reputation mechanisms. This is because the MSN applications should be designed in a distributed and privacy-preserving manner where the identities and long-term relations should not be disclosed. The TSE systems can work well in the MSN where most users behave honestly. Nevertheless, as mentioned in the previous section, if most users behave dishonestly, their behavior will damage the group reputation in a long run and the reputation systems will come into play. The existing open problems in the trustworthy service evaluation research are

- Explore social relation to improve the effectiveness of service dissemination. The service dissemination in the TSE is as important as the review submission. The vendors aim to deliver the service information to the customers in an efficient way. It is a useless dissemination if the customers are not interested in the received service information. Enabling users to explore their social relationships to selectively disseminate the service information could be one promising solution. Further investigation on this research direction is needed.
- Study on the incentive mechanism for the user cooperation. In the TSE system, due to the lack of any trusted authority, the attacks by either vendors or users are hard to prevent. We show that the user cooperation can effectively mitigate

the mentioned review attacks. However, in reality, users are selfish; without any rewards, they hardly participate in the cooperative review submission task. One possible solution is to build a incentive mechanism in the TSE system to reward the users who are cooperative and punish the users who are non-cooperative.

References

1. M. Stanley, Tablet demand and disruption mobile users come of age, Tech. Rep., 2011
2. K. Puttaswamy, S. Wang, T. Steinbauer, D. Agrawal, A. El Abbadi, C. Kruegel, B. Zhao, Preserving location privacy in geo-social applications. IEEE Trans. Mob. Comput. (2012), http://ieeexplore.ieee.org/xpl/articleDetails.jsp?tp=&arnumber=6365638
3. H. Li, H. Hu, J. Xu, Nearby friend alert: location anonymity in mobile geo-social networks. IEEE Pervasive Comput. (2012), http://ieeexplore.ieee.org/xpl/articleDetails.jsp?tp=&arnumber=6392162
4. M. Motani, V. Srinivasan, P. Nuggehalli, Peoplenet: engineering a wireless virtual social network, in *MOBICOM* (ACM, 2005), pp. 243–257
5. X. Liang, X. Li, R. Lu, X. Lin, X. Shen, Seer: a secure and efficient service review system for service-oriented mobile social networks, in *International Conference on Distributed Computing Systems* (IEEE, 2012), pp. 647–656
6. M. Brereton, P. Roe, M. Foth, J.M. Bunker, L. Buys, Designing participation in agile ridesharing with mobile social software, in *OZCHI* (2009), pp. 257–260
7. Vanpool market action plan. Victoria Transport Policy Institute, Tech. Rep., 2003
8. M. Conti, S. Giordano, Multihop ad hoc networking: the theory. IEEE Commun. Mag. **45**(4), 78–86 (2007)
9. K. Viswanath, K. Obraczka, G. Tsudik, Exploring mesh and tree-based multicast routing protocols for manets. IEEE Trans. Mob. Comput. **5**(1), 28–42 (2006)
10. K.M.E. Defrawy, G. Tsudik, Alarm: anonymous location-aided routing in suspicious manets. IEEE Trans. Mob. Comput. **10**(9), 1345–1358 (2011)
11. K. Sohrabi, J. Gao, V. Ailawadhi, G. Pottie, Protocols for self-organization of a wireless sensor network. IEEE Pers. Commun. **7**(5), 16–27 (2000)
12. X. Li, H. Frey, N. Santoro, I. Stojmenovic, Strictly localized sensor self-deployment for optimal focused coverage. IEEE Trans. Mob. Comput. **10**(11), 1520–1533 (2011)
13. X. Li, I. Lille, R. Falcón, A. Nayak, I. Stojmenovic, Servicing wireless sensor networks by mobile robots. IEEE Commun. Mag. **50**(7), 147–154 (2012)
14. K.R. Fall, A delay-tolerant network architecture for challenged internets, in *Special Interest Group on Data Communication* (ACM, 2003), pp. 27–34
15. H. Zhu, X. Lin, R. Lu, Y. Fan, X. Shen, Smart: a secure multilayer credit-based incentive scheme for delay-tolerant networks. IEEE Trans. Veh. Technol. **58**(8), 4628–4639 (2009)
16. Q. Li, S. Zhu, G. Cao, Routing in socially selfish delay tolerant networks, in *Proceedings of IEEE INFOCOM* (IEEE, 2010), pp. 857–865
17. R. Lu, X. Lin, X. Shen, Spring: a social-based privacy-preserving packet forwarding protocol for vehicular delay tolerant networks, in *Proceedings of IEEE INFOCOM* (IEEE, 2010), pp. 632–640

X. Liang et al., *Security and Privacy in Mobile Social Networks*, SpringerBriefs in Computer Science, DOI 10.1007/978-1-4614-8857-6, © The Author(s) 2013

18. F. Li, Y. Wang, Routing in vehicular ad hoc networks: a survey. IEEE Veh. Technol. Mag.
    2(2), 12–22 (2007)
19. M. Raya, J.-P. Hubaux, Securing vehicular ad hoc networks. J. Comput. Secur. **15**(1), 39–68
    (2007)
20. A. Wasef, R. Lu, X. Lin, X. Shen, Complementing public key infrastructure to secure
    vehicular ad hoc networks. IEEE Wirel. Commun. **17**(5), 22–28 (2010)
21. R. Lu, X. Lin, H. Luan, X. Liang, X. Shen, Pseudonym changing at social spots: an effective
    strategy for location privacy in vanets. IEEE Trans. Veh. Technol. **61**(1), 86–96 (2011)
22. W. Dong, V. Dave, L. Qiu, Y. Zhang, Secure friend discovery in mobile social networks, in
    Proceedings of IEEE INFOCOM (IEEE, 2011), pp. 1647–1655
23. G. Chen, F. Rahman, Analyzing privacy designs of mobile social networking applications, in
    IEEE/IFIP International Conference on Embedded and Ubiquitous Computing, vol. 2 (IEEE,
    2008), pp. 83–88
24. M. Li, N. Cao, S. Yu, W. Lou, Findu: privacy-preserving personal profile matching in mobile
    social networks, in *Proceedings of IEEE INFOCOM* (IEEE, 2011), pp. 2435–2443
25. X. Liang, X. Li, T.H. Luan, R. Lu, X. Lin, X. Shen, Morality-driven data forwarding with
    privacy preservation in mobile social networks. IEEE Trans. Veh. Technol. **7**(61), 3209–3222
    (2012)
26. S. Gaonkar, J. Li, R.R. Choudhury, L.P. Cox, A. Schmidt, Micro-blog: sharing and querying
    content through mobile phones and social participation, in *Mobile Systems, Applications, and
    Services* (ACM, 2008), pp. 174–186
27. R. Lu, X. Lin, X. Liang, X. Shen, A secure handshake scheme with symptoms-matching for
    mhealthcare social network. ACM Mob. Netw. Appl. (MONET) **16**(6), 683–694 (2011)
28. R. Zhang, Y. Zhang, J. Sun, G. Yan, Fine-grained private matching for proximity-based
    mobile social networking, in *Proceedings of IEEE INFOCOM* (IEEE, 2012), pp. 1969–1977
29. Z. Yang, B. Zhang, J. Dai, A.C. Champion, D. Xuan, D. Li, E-smalltalker: a distributed
    mobile system for social networking in physical proximity, in *International Conference on
    Distributed Computing Systems* (IEEE, 2010), pp. 468–477
30. K. Lee, S. Hong, S.J. Kim, I. Rhee, S. Chong, Slaw: a new mobility model for human walks,
    in *Proceedings of IEEE INFOCOM* (IEEE, 2009), pp. 855–863
31. I. Rhee, M. Shin, S. Hong, K. Lee, S.J. Kim, S. Chong, On the levy-walk nature of human
    mobility. IEEE/ACM Trans. Netw. **19**(3), 630–643 (2011)
32. J. Scott, R. Gass, J. Crowcroft, P. Hui, C. Diot, A. Chaintreau, CRAWDAD trace cam-
    bridge/haggle/imote/infocom2006 (v. 2009-05-29), 2009
33. N. Eagle, A. Pentland, Reality mining: sensing complex social systems. Pers. Ubiquitous
    Comput. **10**(4), 255–268 (2006)
34. J. Yeo, D. Kotz, T. Henderson, Crawdad: a community resource for archiving wireless data at
    dartmouth. Comput. Commun. Rev. **36**(2), 21–22 (2006)
35. F. Fukuyama, *Trust: Social Virtues and the Creation of Prosperity* (Free Press, New York,
    1995)
36. A. Colman, Cooperation, psychological game theory, and limitations of rationality in social
    interaction. Behav. Brain Sci. **26**(02), 139–153 (2003)
37. M. Wubben, Social Functions of Emotions in Social Dilemmas, Rotterdam, 2010
38. T. Ketelaar, W.T. Au, The effects of feelings of guilt on the behaviour of uncooperative
    individuals in repeated social bargaining games: an effect-as-information interpretation of
    the role of emotion in social interaction. Cogn. Emot. **17**(3), 429–453 (2003)
39. W. Wang, X.-Y. Li, Y. Wang, Truthful multicast routing in selfish wireless networks, in
    MobiCom (ACM, 2004), pp. 245–259
40. Z. Li, H. Shen, Game-theoretic analysis of cooperation incentive strategies in mobile ad hoc
    networks. IEEE Trans. Mob. Comput. **11**(8), 1287–1303 (2012)
41. D. McMillan, D. Chavis, Sense of community: a definition and theory. J. Community Psychol.
    14(1), 6–23 (1986)
42. D. Perkins, P. Florin, R. Rich, A. Wandersman, D. Chavis, Participation and the social and
    physical environment of residential blocks: crime and community context. Am. J. Community
    Psychol. **18**(1), 83–115 (1990)

43. S. Okasha, Altruism, group selection and correlated interaction. Br. J. Philos. Sci. **56**(4), 703–725 (2005)

44. A. Mei, G. Morabito, P. Santi, J. Stefa, Social-aware stateless forwarding in pocket switched networks, in *Proceedings of IEEE INFOCOM* (IEEE, 2011), pp. 251–255

45. E.M. Daly, M. Haahr, Social network analysis for routing in disconnected delay-tolerant manets, in *MobiHoc* (ACM, 2007), pp. 32–40

46. P. Hui, J. Crowcroft, E. Yoneki, Bubble rap: social-based forwarding in delay tolerant networks, in *MobiHoc* (ACM, 2008), pp. 241–250

47. J. Fan, Y. Du, W. Gao, J. Chen, Y. Sun, Geography-aware active data dissemination in mobile social networks, in *Mobile Adhoc and Sensor Systems* (IEEE, 2010), pp. 109–118

48. P. Hui, E. Yoneki, S.Y. Chan, J. Crowcroft, Distributed community detection in delay tolerant networks, in *Proceedings of 2nd ACM/IEEE International Workshop on Mobility in the Evolving Internet Architecture*, ser. MobiArch (ACM, New York, 2007), pp. 1–8

49. N.P. Nguyen, T.N. Dinh, S. Tokala, M.T. Thai, Overlapping communities in dynamic networks: their detection and mobile applications, in *MOBICOM* (ACM MobiCom, 2011), pp. 85–96

50. P. Hui, E. Yoneki, J. Crowcroft, S.Y. Chan, Identifying social communities in complex communications for network efficiency, in Complex Sciences Lecture Notes of the Institute for Computer Sciences, Social Informatics and Telecommunications Engineering **4**, (2009), pp. 351–363

51. P.V. Marsden, Egocentric and sociocentric measures of network centrality. Complex (1) **24**(4), 407–422 (2002)

52. D. Pointcheval, J. Stern, Security proofs for signature schemes, in *EUROCRYPT* (1996), pp. 387–398

53. X. Boyen, B. Waters, Full-domain subgroup hiding and constant-size group signatures, Public Key Cryptography – PKC 2007 Lecture Notes in Computer Science **4450**, (2007), pp. 1–15

54. X. Liang, Z. Cao, J. Shao, H. Lin, Short group signature without random oracles, in *International Conference on Information and Communications Security* Lecture Notes in Computer Science **4861**, (2007), pp. 69–82

55. X. Liang, X. Li, K. Zhang, R. Lu, X. Lin, X. Shen, Fully anonymous profile matching in mobile social networks. IEEE J. Sel. Areas Commun. **31**(9), 641–655 (2013)

56. B. Viswanath, A. Post, P.K. Gummadi, A. Mislove, An analysis of social network-based sybil defenses, in *Special Interest Group on Data Communication* (ACM, 2010), pp. 363–374

57. A. Mohaisen, N. Hopper, Y. Kim, Keep your friends close: incorporating trust into social network-based sybil defenses, in *Proceedings of IEEE INFOCOM* (IEEE, 2011), pp. 1943–1951

58. W. Wei, F. Xu, C.C. Tan, Q. Li, Sybildefender: defend against sybil attacks in large social networks, in *Proceedings of IEEE INFOCOM* (IEEE, 2012), pp. 1951–1959

59. G. Wang, M. Mohanlal, C. Wilson, X. Wang, M.J. Metzger, H. Zheng, B.Y. Zhao, Social turing tests: crowdsourcing sybil detection. in *CoRR*, vol. abs/1205.3856 (2012)

60. X. Liang, X. Lin, X. Shen, Enabling trustworthy service evaluation in service-oriented mobile social networks. IEEE Trans. Parallel Distrib. Syst. (2013), ieeexplore.ieee.org/xpl/articleDetails.jsp?tp=&arnumber=6463403

61. R. Gross, A. Acquisti, H.J. Heinz III, Information revelation and privacy in online social networks, in *Workshop on Privacy in the Electronic Society* (ACM, 2005), pp. 71–80

62. F. Stutzman, An evaluation of identity-sharing behavior in social network communities. iDMAa J. **3**(1), 10–18 (2006)

63. K.P.N. Puttaswamy, A. Sala, B.Y. Zhao, Starclique: guaranteeing user privacy in social networks against intersection attacks, in *Conference on emerging Networking EXperiments and Technologies* (ACM, 2009), pp. 157–168

64. E. Zheleva, L. Getoor, To join or not to join: the illusion of privacy in social networks with mixed public and private user profiles, in *World Wide Web* (ACM, 2009), pp. 531–540

65. B. Falchuk, S. Loeb, Privacy enhancements for mobile and social uses of consumer electronics. IEEE Commun. Mag. **48**(6), 102–108 (2010)

66. F. Cuadrado, J.C. Dueñas, Mobile application stores: success factors, existing approaches, and future developments. IEEE Commun. Mag. **50**(11), 160–167 (2012)
67. D. Balfanz, G. Durfee, N. Shankar, D.K. Smetters, J. Staddon, H.-C. Wong, Secret handshakes from pairing-based key agreements, in *IEEE Symposium on Security and Privacy* (IEEE, 2003), pp. 180–196
68. M.J. Freedman, K. Nissim, B. Pinkas, Efficient private matching and set intersection, in *Advances in Cryptology-EUROCRYPT* Lecture Notes in Computer Science **3027**, (2004), pp. 1–19
69. M. Jakobsson, J.-P. Hubaux, L. Buttyan, A micro-payment scheme encouraging collaboration in multi-hop cellular networks, in *Financial Cryptography* Lecture Notes in Computer Science **2742**, (2003), pp. 15–33
70. V. Srinivasan, P. Nuggehalli, C.-F. Chiasserini, R. R. Rao, An analytical approach to the study of cooperation in wireless ad hoc networks. IEEE Trans. Wirel. Commun. **4**(2), 722–733 (2005)
71. S. Zhong, E.L. Li, Y.G. Liu, Y.R. Yang, On designing incentive-compatible routing and forwarding protocols in wireless ad-hoc networks: an integrated approach using game theoretical and cryptographic techniques, in *MOBICOM* (ACM Mobicom, 2005), pp. 117–131
72. L. Sweeney, k-anonymity: a model for protecting privacy. Int. J. Uncertain. Fuzziness Knowl. Based Syst. **10**(5), 557–570 (2002)
73. L. von Ahn, A. Bortz, N.J. Hopper, k-anonymous message transmission, in *ACM Conference on Computer and Communications Security* (ACM, 2003), pp. 122–130
74. P. Wang, P. Ning, D.S. Reeves, A *k*-anonymous communication protocol for overlay networks, in *ACM Symposium on Information, Computer and Communications Security* (ACM, 2007), pp. 45–56
75. J. Freudiger, M.H. Manshaei, J.-P. Hubaux, D.C. Parkes, On non-cooperative location privacy: a game-theoretic analysis, in *ACM Conference on Computer and Communications Security* (ACM, 2009), pp. 324–337
76. G. Box, G.M. Jenkins, G.C. Reinsel, *Time Series Analysis: Forecasting and Control*, 4th edn. (Wiley, New York, 2008)
77. X. Li, N. Mitton, D. Simplot-Ryl, Mobility prediction based neighborhood discovery for mobile ad hoc networks **6640**, (2011), pp. 241–253
78. D. Boneh, M.K. Franklin, Identity-based encryption from the weil pairing, in *CRYPTO* (2001), pp. 213–229
79. F. Hess, Efficient identity based signature schemes based on pairings, in *Selected Areas in Cryptography* Lecture Notes in Computer Science **2595**, (2003), pp. 310–324
80. C. Gentry, Z. Ramzan, Identity-based aggregate signatures, in *Public Key Cryptography* Lecture Notes in Computer Science **3958**, (2006), pp. 257–273
81. A. Shamir, How to share a secret. Commun. ACM **22**(11), 612–613 (1979)
82. P. Paillier, Public-key cryptosystems based on composite degree residuosity classes, in *Advances in Cryptology-EUROCRYPT 99* Lecture Notes in Computer Science **1592**, (1999), pp. 223–238
83. M. Naehrig, K. Lauter, V. Vaikuntanathan, Can homomorphic encryption be practical? in *Cloud Computing Security Workshop* (ACM, 2011), pp. 113–124
84. R. Lu, X. Liang, X. Li, X. Lin, X. Shen, Eppa: an efficient and privacy-preserving aggregation scheme for secure smart grid communications. IEEE Trans. Parallel Distrib. Syst. **23**(9), 1621–1631 (2012)
85. I.F. Blake, V. Kolesnikov, Strong conditional oblivious transfer and computing on intervals, in *ASIACRYPT* (ACM, 2004), pp. 515–529
86. J. Katz, A. Sahai, B. Waters, Predicate encryption supporting disjunctions, polynomial equations, and inner products, in J. of Cryp. **26**(2), 191–224 (2013)
87. D.J. Watts, Small worlds: the dynamics of networks between order and randomness. J. Artif. Soc. Soc. Simul. **7**(2), (2004), http://jasss.soc.surrey.ac.uk/index_by_issue.html
88. C. Bron, J. Kerbosch, Finding all cliques of an undirected graph (algorithm 457). Commun. ACM **16**(9), 575–576 (1973)

89. L. Kissner, D.X. Song, Privacy-preserving set operations, in *Advances in Cryptology–CRYPTO* Lecture Notes in Computer Science **3621**, (2005), pp. 241–257
90. Q. Ye, H. Wang, J. Pieprzyk, Distributed private matching and set operations, in *Information Security Practice and Experience* Lecture Notes in Computer Science **4991**, (2008), pp. 347–360
91. D. Dachman-Soled, T. Malkin, M. Raykova, M. Yung, Efficient robust private set intersection, in *Applied Cryptography and Network Security* Lecture Notes in Computer Science **5536**, (2009), pp. 125–142
92. S. Jarecki, X. Liu, Efficient oblivious pseudorandom function with applications to adaptive OT and secure computation of set intersection, in *Theory of Cryptography* Lecture Notes in Computer Science **5444**, (2009), pp. 577–594
93. C. Hazay, Y. Lindell, Efficient protocols for set intersection and pattern matching with security against malicious and covert adversaries. J. Cryptol. **23**(3), 422–456 (2010)
94. B. Goethals, S. Laur, H. Lipmaa, T. Mielikäinen, On private scalar product computation for privacy-preserving data mining, in *Information Security and Cryptology–ICISC 2004* Lecture Notes in Computer Science **3506**, (2005), pp. 104–120
95. A.C.-C. Yao, Protocols for secure computations (extended abstract), in *Foundations of Computer Science* (IEEE, 1982), pp. 160–164
96. O. Goldreich, S. Micali, A. Wigderson, How to play any mental game or a completeness theorem for protocols with honest majority, in *Symposium on Theory of Computing* (ACM, 1987), pp. 218–229
97. I. Ioannidis, A. Grama, M.J. Atallah, A secure protocol for computing dot-products in clustered and distributed environments, in *International Conference on Parallel Processing* (IEEE, 2002), pp. 379–384
98. Q. Yuan, I. Cardei, J. Wu, Predict and relay: an efficient routing in disruption-tolerant networks, in *MobiHoc* (ACM, 2009), pp. 95–104
99. X. Lin, R. Lu, X. Liang, X. Shen, Stap: a social-tier-assisted packet forwarding protocol for achieving receiver-location privacy preservation in vanets, in *Proceedings of IEEE INFOCOM* (IEEE, 2011), pp. 2147–2155
100. H. Luan, L. Cai, J. Chen, X. Shen, F. Bai, Vtube: towards the media rich city life with autonomous vehicular content distribution, in *Sensor, Mesh and Ad Hoc Communications and Networks* (2011), pp. 359–367
101. A. Lindgren, A. Doria, O. Schelén, Probabilistic routing in intermittently connected networks, in *ACM SIGMOBILE Mobile Computing and Communications Review* **7**(3), 19–20 (2003)
102. W. Gao, Q. Li, B. Zhao, G. Cao, Multicasting in delay tolerant networks: a social network perspective, in *MobiHoc* (ACM, 2009), pp. 299–308
103. P. Hui, J. Crowcroft, E. Yoneki, Bubble rap: social-based forwarding in delay-tolerant networks. IEEE Trans. Mob. Comput. **10**(11), 1576–1589 (2011)
104. W. Yu, K.J.R. Liu, Game theoretic analysis of cooperation stimulation and security in autonomous mobile ad hoc networks. IEEE Trans. Mob. Comput. **6**(5), 507–521 (2007).
105. H. Zhu, X. Lin, R. Lu, Y. Fan, X. Shen, Smart: a secure multilayer credit-based incentive scheme for delay-tolerant networks. IEEE Trans. Veh. Technol. **58**(8), 4628–4639 (2009)
106. M.H. Manshaei, Q. Zhu, T. Alpcan, T. Basar, J.-P. Hubaux, Game theory meets network security and privacy. Ecole Polytechnique Fédérale de Lausanne (EPFL), Tech. Rep., September 2010, epfl-report-151965
107. M. Raya, R. Shokri, J.-P. Hubaux, On the tradeoff between trust and privacy in wireless ad hoc networks, in *WISEC* (ACM, 2010), pp. 75–80
108. M. Mahmoud, X. Shen, Pis: a practical incentive system for multihop wireless networks. IEEE Trans. Veh. Technol. **59**(8), 4012–4025 (2010)
109. Z. Li, H. Shen, Game-theoretic analysis of cooperation incentive strategies in mobile ad hoc networks. IEEE Trans. Mob. Comput. **11**(8), 1287–1303 (2012)
110. C.Y.T. Ma, D.K.Y. Yau, N.K. Yip, N.S.V. Rao, Privacy vulnerability of published anonymous mobility traces, in *MobiCom* (ACM, 2010), pp. 185–196

111. M. Li, K. Sampigethaya, L. Huang, R. Poovendran, Swing & swap: user-centric approaches towards maximizing location privacy, in *Workshop on Privacy in the Electronic Society* (ACM, 2006), pp. 19–28
112. D. Chaum, Untraceable electronic mail, return addresses, and digital pseudonyms. Commun. ACM **24**(2), 84–88 (1981)
113. K. Hoffman, D. Zage, C. Nita-Rotaru, A survey of attack and defense techniques for reputation systems. ACM Comput. Surv. **42**(1) (2009)
114. R. Lu, X. Lin, H. Zhu, X. Shen, B. Preiss, Pi: a practical incentive protocol for delay tolerant networks. IEEE Trans. Wirel. Commun. **9**(4), 1483–1493 (2010)
115. Social group, Wikipedia, http://en.wikipedia.org/wiki/Social_group
116. B. Waters, Efficient identity-based encryption without random oracles, in *Advances in Cryptology-EUROCRYPT* Lecture Notes in Computer Science **3494**, (2005), pp. 114–127
117. D. Boneh, X. Boyen, Short signatures without random oracles, in *Advances in Cryptology-EUROCRYPT* Lecture Notes in Computer Science **3027**, (2004), pp. 56–73
118. I. Wen, Factors affecting the online travel buying decision: a review. Int. J. Contemp. Hosp. Manag. **21**(6), 752–765 (2009)
119. S. Dhar, U. Varshney, Challenges and business models for mobile location-based services and advertising. Commun. ACM **54**(5), 121–128 (2011)
120. H. Rajan, M. Hosamani, Tisa: toward trustworthy services in a service-oriented architecture. IEEE Trans. Serv. Comput. **1**(4), 201–213 (2008)
121. Y. Wang, L. Li, Two-dimensional trust rating aggregations in service-oriented applications. IEEE Trans. Serv. Comput. **4**(4), 257–271 (2011)
122. E. Ayday, F. Fekri, Iterative trust and reputation management using belief propagation. IEEE Trans. Dependable Secure Comput. **9**(3), 375–386 (2012)
123. A. Das, M.M. Islam, Securedtrust: a dynamic trust computation model for secured communication in multiagent systems. IEEE Trans. Dependable Secure Comput. **9**(2), 261–274 (2012)
124. J. Douceur, The sybil attack, in *Peer-to-Peer Systems* Lecture Notes in Computer Science **2429**, (2002), pp. 251–260
125. C. Karlof, D. Wagner, Secure routing in wireless sensor networks: attacks and countermeasures. Ad Hoc Netw. **1**(2–3), 293–315 (2003)
126. J. Newsome, E. Shi, D.X. Song, A. Perrig, The sybil attack in sensor networks: analysis & defenses, in ACM *Proceedings of the 3rd international symposium on Information processing in sensor networks* (2004), pp. 259–268
127. R. Lu, X. Lin, X. Liang, X. Shen, A dynamic privacy-preserving key management scheme for location-based services in vanets. IEEE Trans. Intell. Transp. Syst. **13**(1), 127–139 (2012)